Proceedings of the Gökova Geometry-Topology Conference 2024

Editors

Selman Akbulut
Gökova Geometry Topology Institute, Muğla, Turkey

Denis Auroux
Department of Mathematics, Harvard University, Cambridge, Massachusetts, U.S.A.

Turgut Önder
Department of Mathematics, Middle East Technical University
Ankara, Turkey

Proceedings of the Gökova Geometry-Topology Conference 2024

Editors:
Selman Akbulut, Gökova Geometry Topology Institute, Muğla, Turkey
Denis Auroux, Harvard University, Cambridge, Mass., U.S.A.
Turgut Önder, Middle East Technical University, Ankara, Turkey

Published in 2025 by International Press, Somerville, Massachusetts, U.S.A. under license from Gökova Geometry-Toplogy Conferences.

ISBN (print): 978-1-57146-583-2
ISBN (online): 978-1-57146-584-9

Printed July 2025 in the United States of America.

29th GÖKOVA GEOMETRY-TOPOLOGY CONFERENCE

May 27 – May 31, 2024

Participants of the 29th Gökova Geometry-Topology Conference

PREFACE

This volume contains articles from the lectures of the 29th Gökova Geometry-Topology Conference, held on the Gökova Bay from May 27 to May 31, 2024. It contains three inspiring research articles and a note about "corks". The articles are on interesting topics: "Tropicalization and lines on surfaces", "Algebraic structures on symplectic cohomology", and on "Symplectic flexible links". The 29th Gökova Geometry-Topology Conference was partially supported by the Turkish Mathematical Society, Yücelen Group, and "Santori Family Charitable Foundation". Lastly, we want to thank Eylem Zeliha Yıldız for compiling this PGGT volume.

May 6, 2025

The Editors

CONTENTS

PSL_2 tropicalization and lines on surfaces 1
M. SHKOLNIKOV and P. PETROV

Descent with algebraic structures for symplectic cohomology 18
U. VAROLGUNES

Construction of symplectic flexible links 37
J. BJÖRKLUND and G. DIMITROGLOU RIZELLL

Corks 57
S. AKBULUT

Proceedings of 29th Gökova
Geometry-Topology Conference
pp. 1 – 17

PSL_2 tropicalization and lines on surfaces

Mikhail Shkolnikov, Peter Petrov

ABSTRACT. The paper is based on a talk given by the first author at the Gökova Geometry & Topology conference in May 2024. The subject is an interplay between the ideas of tropical geometry and two-by-two matrices with an intention to explore new types of geometries. More concretely, the article gives a preliminary account for a non-abelian version of phase tropicalization for subvarieties of PSL_2.

1. Introduction

Tropical geometry is concerned with polyhedral spaces endowed with an integral affine structure and relations between them [1]. Some authors [2] choose to say that "tropical geometry is a combinatorial shadow of algebraic geometry". Indeed, in a number of setups, algebraic varieties (or their families) may cast a shadow that will have a tropical shape. Such a procedure is referred to as tropicalization. In the present article, we continue the study of similar type procedure initiated in [3] and recently refined in [4].

We would like to remark that the quote above is not entirely accurate. First, there are complete tropical spaces that mimic the structure of non-algebraic complex manifolds, with most basic examples being generic tori and Hopf manifolds [5]. Second, there are symplectic aspects of the theory, allowing for example, to construct Lagrangian submanifolds starting with tropical curves [6] or tropical hypersurfaces [7]. Moreover, Gross-Siebert program puts tropical geometry at the core of the fundamental interplay between complex and symplectic geometries [8], namely the Mirror symmetry.

On the other hand, it is certainly the straightest path that most introductions (such as the excellent "Brief introduction to tropical geometry" [9]) take, to imitate the spirit of classical algebraic geometry, where at least initially, one is interested in solutions of systems of polynomial equations, i.e. algebraic varieties. Speaking of solutions, we generally assume some arithmetic system, such as complex numbers, or some other ring or field. In tropical arithmetic, the numbers $\mathbb{T}$ are just real numbers together with $-\infty$, which is the neutral element with respect to tropical addition, replaced now by the operation of taking the maximum. The role of multiplication is taken by the ordinary addition. These operations are degenerations of the classical ones by the procedure known as Litvinov-Maslov dequantization.

Key words and phrases. Phase tropicalization, hyperbolic 3-space, complex surfaces.
M.S. was supported by the Simons Foundation International grant no. 992227, IMI-BAS.

Therefore, a tropical polynomial is a kind of convex piece-wise linear function. The notion of a root is also modified. Namely, we say that a tropical polynomial has a root at some point if the function is not smooth at it. This definition is best explained by redefining the operation of tropical addition to be multi-valued:

$$“\alpha+\beta” = \begin{cases} \max(\alpha,\beta), & \alpha \neq \beta \\ [-\infty,\alpha], & \alpha=\beta. \end{cases}$$

With such an operation tropical numbers form what is known as a hyperfield [10, 11] and the notion of zero locus of a polynomial is now the same as in classical algebraic geometry, i.e. the set of points where the polynomial attains the addition's neutral element.

The motivation behind the above multi-valued refinement of the tropical addition can be explained as follows. Assume we have a real parameter $t \to \infty$ and take two families of complex numbers $A(t)$ and $B(t)$ such that the first asymptotic of each is known, i.e. $A(t) = at^\alpha + o(t^\alpha)$ and $B(t) = bt^\beta + o(t^\beta)$. Then, the limit of $\log_t |A(t) + B(t)|$ is $\max(\alpha, \beta)$ if $\alpha \neq \beta$, but can be anything less or equal otherwise. Taking such limits is the first instance of a tropicalization.

Remarkably, a tropical polynomial of degree $d > 0$ in one variable always has d roots counted with multiplicities, i.e. the fundamental theorem of algebra holds in tropical arithmetic, so $\mathbb{T}$ is algebraically closed. If a tropical Laurent polynomial in two variables is given, it defines a tropical curve C in $(\mathbb{T}^*)^2 = \mathbb{R}^2$. From the geometric perspective, C is a planar graph with straight edges of rational slope equipped with integer weights, satisfying the balancing condition at every vertex. Many classical facts hold for tropical curves. For example, Bernstein's theorem on the number of intersections of two planar curves in given deformation classes holds, provided we understand how to define the multiplicity of each intersection point. By understanding how tropical curves arise as limits of complex curves one is able to deduce analogous fact in algebraic geometry.

Before passing to the main geometric narrative of the paper, we would like to mention a few applications of tropical mathematics to other fields.

Another important direction of investigation gives a general frame for developing tropical differential equations by tropicalizing the equations and their solutions. It is based on idempotent semirings and an idempotent version of differential algebra [12].

In combinatorics, parametrized versions of some classical algorithms for computing shortest-path trees have been expressed in tropical terms [13].

In statistics, there are models expressing probability distributions as polynomial relations, turning to algebraic varieties, so they could be tropicalized (see [14]). Tropicalized statistical models are fundamental for parametric inference, that is when analyzing the behavior of inference algorithms under the variation of model parameters. Inference is the evaluation of one or more coordinates of a single point on an algebraic variety, either over a field or over tropical semi-field.

For instance, the theoretical formulations of statistical mechanics appear to be naturally suited for tropical expression. For example taking the tropical limit of the Schrödinger equation leads to the classical Hamilton–Jacobi equation, the procedure being known in the mathematical literature as Maslov dequantization (see[15, 16]), which may be reminiscent of the WKB approximation.

In biology, an application of tropical geometry leads to simplifying complex biological models. Its methods exploit a feature of biological systems called multiscaleness [17, 18].

Another area of applications to mention is in neural networks. They are based on the hope that in particular it would shed light on the working of deep neural networks [19].

Many classical results such as the fundamental theorem of algebra, Bézout theorem and Pascal's theorem, have been re-proved in tropical context [20, 21, 22]. Others, like the theorem of 27 lines on a smooth cubic surface, has been modified appropriately when proved in tropical geometry [23]. In a number of cases the tropical approach gives another way to obtain results but in a supposedly simpler way, a consequence of the piecewise linear structure of tropical varieties. An example is given by the celebrated Mikhalkin's correspondence theorem in the context of enumerative geometry [24], which procured a huge boost of activity in the field. By counting tropical curves of a given degree and genus passing through an appropriate number of generic points one recovers the number of complex curves, that has been asked initially. In the tropical count, each curve comes with a multiplicity that is dictated by the combinatorial structure of its vertices. The multiplicity of a tropical curve tells how many curves in the classical count tropicalize to it. This approach has been used successfully for other problems from enumerative geometry, like the theorem of Caporaso-Harris [25, 26, 27], giving at the same time an alternative algorithmic methods for eliminating variables [28].

An important feature of the planar case is that every planar tropical curve is realizable as a tropicalization. This fact extends to tropical hypersurfaces in $\mathbb{R}^n$. However, already for tropical curves in $\mathbb{R}^3$ this is not true. For example, in classical geometry, every spacial cubic curve of genus one belongs to some plane, but there is a tropical spacial cubic of genus one that doesn't belong to any tropical plane. The question of realizability is very subtle, a tropical surface and a tropical curve on it can be realizable separately, but not as a pair as was shown in [29].

The two most conventional versions of tropicalization rely on the concept of amoeba, where the name comes from the famous book [30]. In a more geometric version, one considers the scaling limit of amoebas of subvarieties in the complex algebraic torus [31]. For an algebraic subvariety $V \subset (\mathbb{C}^*)^n$ its amoeba is $\mathrm{Log}(V) \subset \mathbb{R}^n$, where $\mathrm{Log}\colon (\mathbb{C}^*)^n \to \mathbb{R}^n$ sends $(z_1, \dots, z_n)$ to $(\log|z_1|, \dots, \log|z_n|)$. Amoebas have many remarkable properties that we are not going to discuss here. By replacing $\mathbb{C}$ in the previous sentence with a field $\mathbb{K}$ endowed with an ultra-norm $|\cdot|_{\mathbb{K}^*}$ (where $-\log \circ |\cdot|_{\mathbb{K}^*}$ is a valuation) we get the notion of a non-Archimedean amoeba, and its closure is called a tropicalization of the subvariety in this approach [32].

In [3] a modified setup was considered. Namely, the map Log was interpreted as taking the quotient of the algebraic torus by its maximal compact subgroup. This interpretation allows a replacement of the torus by some other algebraic group, for example, $SL_2(\mathbb{C})$ or its central quotient $PSL_2(\mathbb{C})$, for which the corresponding amoebas were studied, as well as their tropical limits in the case of families of curves.

What shall be discussed in the next sections, is a modest extension of this work in the case of $PSL_2(\mathbb{C})$. We stick with this particular choice of the group since its natural compactification is the projectivization of the space of complex two-by-two matrices, identified with $\mathbb{C}P^3$. It is slightly more convenient to us than the compactification of $SL_2(\mathbb{C})$, which is a smooth projective quadric three-fold. We are not yet attempting to present an ultimate description of PSL_2 tropical geometry, but rather argue that it requires some decorations in order to capture even the most basic phenomena in the interplay of spatial curves and surfaces.

2. PSL_2 tropicalization

A maximal compact subgroup of the complex algebraic group $PSL_2(\mathbb{C})$ is $PSU(2)$, the quotient of the group $SU(2)$ of unitary matrices with determinant 1 by the subgroup $\{I_2, -I_2\}$, where I_2 denotes the two-by-two identity matrix. Topologically, $SU(2)$ is a three-dimensional sphere. One way to see it is via the identification of $SU(2)$ with the group of quaternions of norm one. More explicitly,

$$SU(2) = \{\begin{pmatrix} z & w \\ -\overline{w} & \overline{z} \end{pmatrix} | z, w \in \mathbb{C}, |z|^2 + |w|^2 = 1\}.$$

The quotient map $SU(2) \to PSU(2)$ is a degree two covering, thus $PSU(2)$ is topologically a three-dimensional projective space.

In fact, it is also true geometrically in the following sense. As it was mentioned, $PSL_2(\mathbb{C})$ is naturally compactified to $\mathbb{C}P^3$, the projectivization of the space of complex two-by-two matrices. Moreover, there exists a real structure on it such that $PSU(2)$ is the corresponding set of real points. Let us construct this real structure explicitly. First, note that the operation of group-theoretical invertion on $PSL_2(\mathbb{C})$ extends to a linear involution on $\mathbb{C}P^3$ given by $[B]_{\mathbb{C}^*} \mapsto [B^{\mathbf{c}}]_{\mathbb{C}^*}$, where $\begin{pmatrix} a & b \\ c & d \end{pmatrix}^{\mathbf{c}} = \begin{pmatrix} d & -b \\ -c & a \end{pmatrix}$ and $[B]_{\mathbb{C}^*} \in \mathbb{C}P^3$ denotes the class of a nonzero matrix up to a rescaling by a non-zero complex number. Then, composing it with the Hermitian conjugation, $[B]_{\mathbb{C}^*} \mapsto [(B^{\mathbf{c}})^*]_{\mathbb{C}^*}$ gives an anti-holomorphic involution on $\mathbb{C}P^3$ with the set of fixed points being $PSU(2)$. Observe also that the quadric surface $Q(\mathbb{C}) = \mathbb{C}P^3 \backslash PSL_2(\mathbb{C})$ given by the homogeneous equation $\det(B) = 0$ is invariant under this involution but doesn't intersect $PSU(2)$. In other words, $Q(\mathbb{C})$ is a real quadric with no real points, such quadrics are also called imaginary.

The analog of the classical logarithmic projection in the theory of amoebas is the quotient map $\varkappa \colon PSL_2(\mathbb{C}) \to PSL_2(\mathbb{C})/PSU(2)$. We call $\varkappa$ the hyperbolic amoeba map since the target is naturally a hyperbolic 3-space $\mathbb{H}^3$, which will now be explained. Note that there is a polar decomposition in the space of complex matrices, where a matrix A is decomposed into a product PU with U being unitary (the angular part) and P being Hermitian (the radial part). If, furthermore, A is non-degenerate we may assume that P is positive definite, and in such case the decomposition is unique. The map $\varkappa$ is seen as forgetting the unitary factor and keeping only the hermitian one.

Thus, our model for the hyperbolic three-space is

$$\mathbb{H}^3 = \{P \in \mathrm{Mat}_{2\times 2}(\mathbb{C}) | P = P^*, P > 0, \det(P) = 1\}.$$

This presentation essentially coincides with the Minkowski model, since the space of two-by-two Hermitian matrices is a real four-dimensional vector space, and the determinant induces on it the quadratic form of signature $(1,3)$. The set $\det(P) = 1$ is a two-sheeted hyperboloid and the condition $P > 0$ of being positively definite chooses one of the two sheets $\mathbb{H}^3$. Note that $-\det$ restricted to $\mathbb{H}^3$ defines a Riemannian metric. More details about this construction can be found in [33].

With this metric, our $\mathbb{H}^3$ is a homogeneous space since $PSL_2(\mathbb{C})$ acts on it isometrically and transitively (in fact, it is the group of all orientation-preserving isometries). Explicitly the action is given by $A(P) = APA^*$, for $A \in SL_2(\mathbb{C})$ which clearly factors through $PSL_2(\mathbb{C})$. Note that our map $\varkappa$ may be now seen as $\varkappa([A]_{\mathbb{C}^*}) = A(I_2)$, where $\det(A) = 1$ is assumed. Such a map is equivalent to taking the quotient by the stabilizer of I_2, which consists of matrices A such that $AA^* = I_2$ – the condition of being unitary. As a byproduct, we just derived the classical isomorphism $PSU(2) \cong SO(3)$, where the latter is a group of rotations around a point in three-space.

Strictly speaking, the Hermitian part in the polar decomposition $A = PU$ is not exactly $\varkappa(A) = AA^*$, but rather its square root since $AA^* = PUU^*P = P^2$. This difference is minor, however, since taking power $h > 0$ of matrices in our model of $\mathbb{H}^3$ defines a homothety $R_h(P) = P^h$ centered in I_2 and with scaling factor h. One way to see it is by observing that the distance from $P \in \mathbb{H}^3$ to I_2 is computed as the absolute value of the logarithm of an eigenvalue of P.

For completeness, we mention the spherical coamoeba map $\varkappa^\circ \colon PSL_2(\mathbb{C}) \to PSU(2)$ which assigns the unitary part of the polar decomposition and forgets the Hermitian one. It will play an important role in describing PSL_2 phase tropicalization. The explicit formula of $\varkappa^\circ$ is the following.

Lemma 2.1. *Let $[A]_{\mathbb{C}^*} \in PSL_2(\mathbb{C})$ such that $\det(A) = 1$. The value of the coamoeba map is expressed as $\varkappa^\circ([A]_{\mathbb{C}^*}) = [A + (A^c)^*]_{\mathbb{R}^*} \in \mathbb{R}P^3 = PSU(2)$.*

Proof. Let $A = PU$ be the polar decomposition. Then, $P = \sqrt{AA^*}$ and $U = P^{-1}A$. We rewrite the square root expression using Cayley-Hamilton theorem that asserts the annihilation of any matrix by its characteristic polynomial. Explicitly, for a two-by-two matrix B we have $\det(B)I_2 - \mathrm{tr}(B)B + B^2 = 0$. Substituting $B = \sqrt{AA^*}$ we get $[P]_{\mathbb{R}^*} = [I_2 + AA^*]_{\mathbb{R}^*}$. Then,

$$\varkappa^\circ([A]_{\mathbb{C}}) = [(I_2 + AA^*)^c A]_{\mathbb{R}^*} = [A + (A^c)^* A^c A]_{\mathbb{R}^*} = [A + (A^c)^*]_{\mathbb{R}^*}. \qquad \square$$

Here is the geometric version of PSL_2 tropicalization of hyperbolic amoebas of curves as it is described in [3]. A hyperbolic amoeba of a subvariety V in $PSL_2(\mathbb{C})$ is $\varkappa(V)$. A scaling sequence is $\{t_\alpha\}_{\{\alpha \in \boldsymbol{A}\}}$, an unbounded subset of real numbers with $t_\alpha > 1$. Then, for a family of degree d reduced irreducible curves $V_\alpha \subset PSL_2\mathbb{C}$ there exists a subsequence V_β such that $R_{t_\beta}(\varkappa(V_\beta)) \subset \mathbb{H}^3$ converges in the Hausdorff sense to an $\mathbb{H}^3$-spherical complex associated with a degree d $\mathbb{H}^3$-floor diagram. The formal definitions of the latter notions are not relevant for the present article so we are not repeating it here. Vaguely speaking, the tropical limits of hyperbolic amoebas of curves are unions of spheres centered in I_2 (floors) and segments of geodesics passing through I_2 (elevators). In the particular case of $d = 1$, there are two options: either there is a single elevator and no floors, or there is a single floor and two elevators.

In [3], the case of PSL_2 tropical limits of surfaces was not covered. This is mostly due to the reason that the result was not expected to be informative. Indeed, even before the limit, if $V \subset PSL_2(\mathbb{C})$ is an odd degree surface, then $\varkappa(V) = \mathbb{H}^3$, so nothing can be extracted from it. The even degree case is slightly less trivial, i.e. $\mathbb{H}^3 \backslash \varkappa(V)$ can be non-empty but it has at most one connected component which is convex. In a separate article, we will show (in the geometric setup of scaling sequences) that a tropical limit of hyperbolic amoebas of even degree surfaces is always a complement to a geometric open ball centered in I_2.

The moral of all these is that bare PSL_2 tropicalization is not good for observing the geometry of surfaces in $\mathbb{P}^3$. For example, on a generic quintic surface there are no lines. But in the hyperbolic-tropical picture, we see infinitely many of their images! The aim of the present article is to show that this affliction has the potential to be cured. We will show that by restoring the phase in PSL_2 tropicalization at least such type of a basic fact becomes visible.

3. Phase tropicalization

The idea of phase tropicalization is that we consider the scaling limit of a family of varieties inside their ambient space, i.e. without forgetting the phase. In the setup of the previous section, it means lifting the h-homotheties $R_h : \mathbb{H}^3 \to \mathbb{H}^3$ to the h-parametric family of diffeomorphisms $\tilde{R}_h : PSL_2(\mathbb{C}) \to PSL_2(\mathbb{C})$ so that $R_h \circ \varkappa = \varkappa \circ \tilde{R}_h$ (similarly, $\varkappa^\circ \circ \tilde{R}_h = \varkappa^\circ$ holds for the coamoeba map), and considering limits of $\tilde{R}_{t_\alpha}(V_\alpha)$ inside $PSL_2(\mathbb{C})$. The lift can be performed using the polar decomposition $\tilde{R}_h(PU) = P^h U$, and it continuously extends to $Q(\mathbb{C})$ as an identity map.

Although it seems to be a triviality at first glance, phase tropicalization of a family of points opens the door for the possibility of computing (or, at least restricting) phase tropical limits of higher dimensional varieties. For example, in the classical situation of the complex algebraic torus consider a family of numbers $z(t) \in \mathbb{C}^*$ parametrized by real $t \to \infty$. Assume that the first asymptotic is known, i.e. $z(t) = ct^\alpha + o(t^\alpha)$ for some $c \in \mathbb{C}^*$ and $\alpha \in \mathbb{R}$. Then $\log_t |z(t)| \to \alpha$, which is the usual tropical limit, we may call the assignment $z(t) \mapsto \alpha$ an ordinary valuation. The phase tropical limit arises if we consider the same type of rescaling but inside $\mathbb{C}^*$, i.e.

$$\frac{z(t)}{|z(t)|}|z(t)|^{(\log(t))^{-1}} \xrightarrow[t\to\infty]{} \exp(\alpha + i \operatorname{Arg}(c)).$$

The assignment to $z(t)$ of the expression on the right is the phase valuation. Applied coordinate-wise, to each point of a subvariety of a higher dimensional algebraic torus, this gives a classical phase-tropical limit.

In the previous sentence, we meant implicitly that our subvariety that we wish to tropicalize is now defined over a field $\mathbb{K}$ whose elements depend on a parameter t. There are various choices for such a field and it is not our intention to discuss all the technical issues specific to different choices. The intuition, however, is as follows. When a subvariety of a torus is defined by a system of equations with coefficients depending on the parameter t it may be treated either as a family of varieties V_t depending on t or as a single variety V over $\mathbb{K}$. In the former interpretation, we may consider the (phase) tropicalization of this family as a limit in $t \to \infty$, and in the latter one, as a closure of the image of a single variety V under the coordinate-wise (phase) valuation. It is generally assumed that the two approaches agree, however, we are not able to suggest a concrete reference where the fact is proven in full generality.

Coming back to matrices, assume that $[A(t)]_{\mathbb{C}^*} \in PSL_2(\mathbb{C})$ and $A(t) = t^\alpha B + o(t^\alpha)$ as $t \to \infty$. Note that if $\det(A(t)) = 1$ then $\alpha \geq 0$. The classical case suggests what should be the scaling h in $\tilde{R}_h$ for $\tilde{R}_{h(t)}([A(t)]_{\mathbb{C}^*})$ to converge, namely we take $h(t) = (\log(t))^{-1}$.

Lemma 3.1. *Consider $A(t) = t^\alpha B + o(t^\alpha)$ as $t \to \infty$. If $\det(A(t)) = 1$, then $\tilde{R}_{(\log(t))^{-1}}([A(t)]_{\mathbb{C}^*})$ converges to $[e^\alpha B + e^{-\alpha}(B^c)^*]_{\mathbb{C}^*} \in PSL_2(\mathbb{C})$.*

Proof. Note that if $\alpha = 0$ and so $\det(B) = 1$, then the limit is just the value of the spherical coamoeba map $\varkappa^\circ$ on B (see Lemma 2.1). Otherwise, if $\alpha > 0$ and so $\det(B) = 0$, we find separately the limits of the unitary and the Hermitian parts in the polar decomposition for $\tilde{R}_{(\log(t))^{-1}}([A(t)]_{\mathbb{C}^*})$. The first one is simply $[B + (B^c)^*]_{\mathbb{R}^*}$ and the second one is $[e^{-\alpha}\operatorname{tr}(BB^*)I_2 + (e^\alpha - e^{-\alpha})BB^*]_{\mathbb{C}^*}$. As an intermediate step for the latter, find the direction at infinity in which $A(t)(A(t))^*$ goes as well as the asymptotic of the distance to I_2 using the relation between the trace to the eigenvalues, and the eigenvalues to the distance. Combining the parts together and using the identities $BB^c = 0$ and $BB^*B = \operatorname{tr}(BB^*)B$ for two-by-two matrices of rank one, we arrive at the desired expression. □

We would like to use the formula derived in Lemma 3.1 as a definition for the new valuation $\mathrm{VAL} : PSL_2(\mathbb{K}) \to PSL_2(\mathbb{C})$. Given $A(t) = t^\alpha B + o(t^\alpha)$ with $\det(A(t)) = 1$, let

$$\mathrm{VAL}([A(t)]_{\mathbb{K}^*}) = [e^\alpha B + e^{-\alpha}(B^c)^*]_{\mathbb{C}^*},$$

be the PSL_2 phase valuation. Note also that if $\det(A(t)) = 0$ then $\tilde{R}_{(\log(t))^{-1}}([A(t)]_{\mathbb{C}^*})$ converges to $[B]_{\mathbb{C}^*} \in Q(\mathbb{C})$, thus we may extend VAL to the map from $\mathbb{K}P^3$ to $\mathbb{C}P^3$.

The formula for VAL of Lemma 3.1, although explicit, is not that convenient to work with. We will use a different perspective on the space $PSL_2(\mathbb{C})$ as well as its compactification $\mathbb{C}P^3$. In order to do it, let's recall the classical one-to-one correspondence between oriented lines in $\mathbb{R}P^3$ and complex points of an imaginary quadric surface. Namely, for every point $p \in Q(\mathbb{C})$ consider its conjugate point q and trace a complex line $l(\mathbb{C})$ through them. By construction, this line is invariant under the real structure, so it makes sense to look at $l(\mathbb{R}) \subset \mathbb{R}P^3$ which cuts $l(\mathbb{C}) \cong \mathbb{C}P^1$ into two discs. Moreover, since $l(\mathbb{C})$ is canonically oriented being a complex manifold, these discs define (opposite) orientations on their common boundary. We pick the orientation on $l(\mathbb{R})$ induced by the disc containing the point p we started with.

As a byproduct of this construction, we define the circle bundle $\mathcal{S}$ on $Q(\mathbb{C})$ where a fiber over a point p is the corresponding real oriented line. We need to write the fiber a bit more explicitly, namely for a point $p = [B]_{\mathbb{C}^*} \in Q(\mathbb{C})$, the fiber $\mathcal{S}_p$ is $[cB + \bar{c}(B^{\mathbf{c}})^*]_{\mathbb{R}^*}$, where c runs through the set of all nonzero complex numbers. This set is canonically identified with classes $[cB]_{\mathbb{R}^*}$. Therefore, we may write the total space of the circle bundle $\mathcal{S}$ simply as the set of all classes $[B]_{\mathbb{R}^*}$, where B runs over all nonzero complex rank one matrices. The projection $\mathcal{S} \to Q(\mathbb{C})$ is just $[B]_{\mathbb{R}^*} \mapsto [B]_{\mathbb{C}^*}$. Then, there is a diffeomorphism $(0,\infty) \times \mathcal{S} \to PSL_2(\mathbb{C})\backslash PSU(2)$ given by $(\alpha, [B]_{\mathbb{R}^*}) \mapsto [e^{\alpha}B + e^{-\alpha}(B^{\mathbf{c}})^*]_{\mathbb{C}^*}$. We complete this presentation by gluing spaces $(0,\infty)\times\mathcal{S}$, $\{0\}\times PSU(2)$ and $\{\infty\}\times Q(\mathbb{C})$.

We call the result the *cone picture* of $\mathbb{C}P^3$, where at the base at infinity there is the imaginary quadric, at the tip there is the real projective space, and in between is the cylinder of the total space of the circle bundle. Note that the fibers of the circle bundle are glued nicely both to the tip and the base. In the first case, the fiber converges to the corresponding line in the real projective space and, in the second case, it contracts to a point on the base. In this picture, the expression of VAL is particularly nice. Namely, for $A(t) = t^{\alpha}B + o(t^{\alpha})$,

$$\mathrm{VAL}([A(t)]_{\mathbb{C}^*}) = \begin{cases} (0, \varkappa^{\circ}(B)), & \det(A(t)) = 1, \det(B) = 1,\ \alpha = 0; \\ (\alpha, [B]_{\mathbb{R}^*}), & \det(A(t)) = 1, \det(B) = 0,\ \alpha > 0; \\ (\infty, [B]_{\mathbb{C}^*}), & \det(A(t)) = 0. \end{cases}$$

4. The case of lines

The goal of this section is to compute the image under VAL for all lines. Starting from now we will be working over the field $\mathbb{K}$ of Puiseux series with complex coefficients and variable $t \to \infty$. To make the formulas a bit less cumbersome, in the statements of Theorems below we have chosen to suppress writing the intersections of the intervals with the set of rational numbers, which are the possible powers of the type of series we are considering. For the sake of simplifying notations, in what follows we suppress the dependence on t for elements of $\mathbb{K}$.

We start with a line L in $\mathbb{K}P^3$ intersecting the quadric $Q(\mathbb{K})$ at a single point $[A]_{\mathbb{K}^*}$. Take any other point $[B]_{\mathbb{K}^*}$ on L. Then, $L\backslash\{[A]_{\mathbb{K}^*}\}$ is parametrized as $[zA + B]_{\mathbb{K}^*}$ by $z \in \mathbb{K}$. Note that $\det(B) \neq 0$, thus we may assume that $\det(B) = 1$ by choosing a different representative in $[B]_{\mathbb{K}^*}$. This assumption implies also that $\det(Az + B) = 1$ for all $z \in \mathbb{K}$. This follows from the fact that $\det(zA + B)$ is, in general, a quadratic polynomial in z with the leading term being $\det(A)z^2$, which vanishes since A is on the quadric. Thus, $\det(zA + B)$ is a linear polynomial. Since L intersects the quadric at a single point, the polynomial has no roots and, therefore, is constant.

Now write A as $A_0t^{\alpha} + o(t^{\alpha})$ and B as $B_0t^{\beta} + o(t^{\beta})$, where A_0 and B_0 are nonzero matrices with complex entries and $\beta \geq 0$ since $\det(B) = 1$. Take $z = ct^{\gamma} + o(t^{\gamma})$, where c is a nonzero complex number. To find the top-order term in $Az + B$, look at the tropical polynomial $\gamma \mapsto \max(\alpha + \gamma, \beta)$, which has a single breaking point at $\gamma_0 = \beta - \alpha \geq 0$ and the minimal value equal to β. Thus, if $\gamma > \gamma_0$ the leading term is $A_0t^{\gamma+\alpha}$, if $\gamma < \gamma_0$ it is Bt^β and if $\gamma = \gamma_0$ these two expressions compete, which might lead to a cancellation, and a lower leading term as a result.

Let us first resolve the latter situation. The cancellation may happen only if A_0 and B_0 are proportional, i.e. if $B_0 = bA_0$ for some $b \in \mathbb{C}$. Take a different parametrization of L by substituting $z = w - bt^{\gamma_0}$ which results in $Az + B = Aw + B'$, where $B' = B - bt^{\gamma_0}A$ has a strictly lower order highest term than B. Therefore, assuming that B in $zA + B$ has the smallest top order term among all such parameterizations of the same line makes this situation impossible, and we may discard this possibility.

Now, assume $\beta > 0$ in which case $\mathrm{VAL}(L)$ doesn't have points over the tip of the cone $\widehat{Q}$. If $\gamma > \gamma_0$, the term $cA_0t^{\gamma+\alpha}$ dominates in $Az + B$. In the cone picture, this gives a ray with base $[A_0]_{\mathbb{C}^*}$ and going between the heights γ_0 and ∞. Since $c \in \mathbb{C}^*$ may vary arbitrarily we get the whole circle in $\mathcal{S}$ over each interior point of the ray in the image of VAL.

At $\gamma = \gamma_0$, the top-order term is $(cA_0 + B_0)t^\beta$. Note that $\det(cA_0 + B_0) = 0$ since we assumed that $\beta > 0$. Thus $c \in \mathbb{C}^* \mapsto [cA_0 + B_0]_{\mathbb{C}^*}$ parametrizes a line l on the quadric $Q(\mathbb{C})$ with two points $[A_0]_{\mathbb{C}^*}$ and $[B_0]_{\mathbb{C}^*}$ missing. Note, however, that for the purpose of computing VAL, we are rather interested in $[cA_0 + B_0]_{\mathbb{R}^*}$ which defines a section of $\mathcal{S}$ restricted to l. It is not possible to continuously extend this section at $[A_0]_{\mathbb{C}^*}$, the base of the above-mentioned ray, but at the other missing point the section does extend by $[B_0]_{\mathbb{R}^*}$, which corresponds precisely to the image of VAL for all the points on L parametrized by z with $\gamma < \gamma_0$.

To complete the description of $\mathrm{VAL}(L)$ for L tangent to $Q(\mathbb{K})$, we are left with the case $\beta = 0$. Again, the ray part corresponding to $\gamma > 0$ and domination of $cAt^{\gamma+\alpha}$ is exactly same as before. What is different is that now we have points over the tip of the cone corresponding to $\gamma \leq -\alpha$. Under VAL, all the points of L parametrized by z with $\gamma < -\alpha$ will be mapped to a single value $\varkappa^\circ[B_0]_{\mathbb{C}^*} = [B_0 + (B_0^*)^{-1}]_{\mathbb{R}^*}$. Finally, if $\gamma = -\alpha$ we get the spherical coamoeba of a complex line parametrized by $c \mapsto [cA_0 + B_0]_{\mathbb{C}^*}$. Since it will not play any role in the discussion below, we simply state that this coamoeba is equal to the plane spanned by the point $\varkappa^\circ[B_0]_{\mathbb{C}^*}$ and the real line corresponding to $[A_0]_{\mathbb{C}^*} \in Q(\mathbb{C})$ with the line removed (here we also implicitly claim that the point cannot belong to the line). To summarize, we have the following.

Theorem 4.1. *Let L be a line in $\mathbb{K}P^3$ intersecting $Q(\mathbb{K})$ in a single point. Then,* $\mathrm{VAL}(L)$ *is either*

$$\{r\} \times \sigma(l \backslash \{p\}) \cup (r, \infty) \times \mathcal{S}|_{\{p\}} \cup \{\infty\} \times \{p\}, \tag{1}$$

where $r > 0$, l is a line on $Q(\mathbb{C})$, $p \in l$ and σ is a section of $\mathcal{S}|_{l\backslash\{p\}}$, or

$$\{0\} \times \varkappa^\circ(l \backslash \{p\}) \cup (0, \infty) \times \mathcal{S}|_{\{p\}} \cup \{\infty\} \times \{p\}, \tag{2}$$

where l is a line on $\mathbb{C}P^3$ tangent to $Q(\mathbb{C})$ at the point p.

Now we proceed with the case of a line L intersecting the quadric $Q(\mathbb{K})$ at two points. Denote such a pair of points by $[A]_{\mathbb{K}^*}$ and $[B]_{\mathbb{K}^*}$, where A and B are nonzero matrices with entries in $\mathbb{K}$ having vanishing determinant. Then, we can parametrize the line L by

$w \mapsto [Aw+B]_{\mathbb{K}^*}$. Again, the vanishing of $\det(A)$ implies that $\det(Aw+B)$ is a linear polynomial. Its only root is at $w=0$ since $\det(B)=0$. Therefore $\det(Aw+B)=uw$ for some $u \in \mathbb{K}^*$. In order to compute VAL, we normalize $Aw+B$ by dividing it by $\sqrt{uw}$. Thus, after appropriately rescaling A and B, we may assume that the line L without the two points on $Q(\mathbb{K})$ is doubly covered by $\mathbb{K}^*$ with the map given by $z \mapsto [Az+Bz^{-1}]_{\mathbb{K}^*}$.

Write $A=A_0t^\alpha+o(t^\alpha)$, $B=B_0t^\beta+o(t^\beta)$ and $z=ct^\gamma+o(t^\gamma)$, where A_0 and B_0 are rank one complex matrices and $c \in \mathbb{C}^*$. To find the leading power of $Az+Bz^{-1}$, we look at the tropical polynomial $\gamma \mapsto \max(\alpha+\gamma, \beta-\gamma)$. Its breaking point, i.e. the tropical root, is at $\gamma_0=\frac{\beta-\alpha}{2}$, with the value at it being $r=\frac{\beta+\alpha}{2}$. Note that since $\det(Az+Bz^{-1})=1$, and r bounds from above the order of the leading term in $Az+Bz^{-1}$ at $\gamma=\gamma_0$, it must be non-negative.

If $r=0$, then for $\gamma=\gamma_0$ no cancellation is possible, since it would further lower the order of $Az+Bz^{-1}$, i.e. A_0 and B_0 are not proportional and thus $p=[A_0]_{\mathbb{C}^*}$ and $q=[B_0]_{\mathbb{C}^*}$ are distinct points on $Q(\mathbb{C})$. Moreover, $1=\det(Az+Bz^{-1})=\det(cA_0+c^{-1}B_0)+o(1)$ for all $c \in \mathbb{C}^*$ implying that the line l passing through p and q doesn't lie on $Q(\mathbb{C})$. The value of VAL for $\gamma=\gamma_0$ is $\varkappa^\circ(l\backslash\{p,q\})$. For $\gamma>\gamma_0$ the leading term in the parametrization is $cA_0t^{\alpha+\gamma}$, which gives a ray going all the way from the tip to p at the base at infinity, and the whole circle fiber of $\mathcal{S}$. Same goes for the case $\gamma<\gamma_0$ with p being replaced by q.

If $r>0$, cancellations are in principle possible, but only when A_0 and B_0 are proportional. Assume for now the contrary, i.e. that they represent two distinct points p and q on $Q(\mathbb{C})$. Note that the line l passing through these two points now lies on the quadric since $\det(cA_0+c^{-1}B_0)=0$ for all $c \in \mathbb{C}^*$ because $1=\det(Az+Bz^{-1})=\det(cA_0+c^{-1}B_0)t^{2r}+o(t^{2r})$ and $2r>0$. Thus, at $\gamma=\gamma_0$ and by varying c we get a continuous section σ of $\mathcal{S}|_{l\backslash\{p,q\}}$ as the image of VAL. Again, for $\gamma \neq \gamma_0$, we get two rays going from p and q until the height r with the whole fiber of $\mathcal{S}$ over each point.

We are left to study the case when the cancellation occurs. As mentioned, this may happen only if $r>0$. By rescaling, we may assume that it happened at $z=1$, in particular the series expansion of matrices A and B starts as $A=A_0t^r+A_1t^{\alpha_1}+o(t^{\alpha_1})$ $B=-A_0t^r+B_1t^{\beta_1}+o(t^{\beta_1})$, where $r>\alpha_1,\beta_1$. Write also $z=1+ut^\delta+o(t^\delta)$, where $\delta<0$, and $z^{-1}=1-ut^\delta+o(t^\delta)$. We have

$$Az+Bz^{-1}=2A_0ut^{r+\delta}+A_1t^{\alpha_1}+B_1t^{\beta_1}+\text{smaller order terms.}$$

Note that since the determinant of this expression is always 1, by substituting $u=0$ we get $r_1:=\max(\alpha_1,\beta_1)\geq 0$. If $r_1=0$, then we get the same type of expression as (2) in Theorem 4.1, because no further cancellation may happen. If $r_1>0$, the ray at heights (r,∞) in the cone picture continues down to the height r_1, where two things may happen. Either there is no cancellation and we get an expression of the type (1), or there is a new cancellation at the top order, and we need to go to a deeper asymptotic expansion. In the latter case, we would have again proportionality of the top matrix to A_0, meaning that the ray continues in a straight way. The procedure will terminate eventually by virtue of working with Puiseux series.

Theorem 4.2. *Let L be a line in $\mathbb{K}P^3$ intersecting $Q(\mathbb{K})$in two points. Then* VAL(L) *is either*

$$\{r\} \times \sigma(l\backslash\{p,q\}) \cup (r,\infty) \times \mathcal{S}|_{\{p,q\}} \cup \{\infty\} \times \{p,q\}, \tag{3}$$

where $r > 0$, l is a line on $Q(\mathbb{C})$, p and q are distinct points on l and σ is a section of $\mathcal{S}|_{l\backslash\{p,q\}}$, or

$$\{0\} \times \varkappa^\circ(l\backslash\{p,q\}) \cup (0,\infty) \times \mathcal{S}|_{\{p,q\}} \cup \{\infty\} \times \{p,q\}, \tag{4}$$

where l is a line in $\mathbb{K}P^3$ intersecting $Q(\mathbb{C})$ at two points p and q, or VAL(L) *is of the form (1) or (2).*

The only type of lines we haven't considered yet is that of L belonging entirely to $Q(\mathbb{K})$. One way to deal with it is to think of such a line as being parametrized by $[v_1 \cdot v_2^\top]_{\mathbb{K}^*}$, where v_1 and v_2 are two nonzero vectors, one of which is fixed and the other varies freely. The corresponding image under VAL will be a line of the same type but now on $Q(\mathbb{C})$ simply by taking the top degree part of the fixed vector – the top degree part of the second vector will again vary freely.

As a final remark for this section, we emphasize that the computations above should be doable whenever there is an explicit parametrization of a curve, such as when the curve is rational. It will certainly be harder, the case of conics already looks intimidating. However, we expect to give a complete description of VAL images for all rational curves. Higher genus case, on the other hand, at the moment looks virtually impossible to handle in full generality.

5. The case of surfaces

In this section, we aim to find restrictions on the image under VAL for some surfaces in every degree as a preparation for the following section where we will see that no line belongs to any of them (assuming that the input complex polynomials f_k are generic) starting with degree 4. By the nature of the setup, there is a slight variation between even and odd degrees reflected in the odd degree by the presence of a spherical coamoeba of a generic plane over the tip of the cone. At this moment, we know what is its shape – a complement to a cylinder; this fact, however, is not relevant to the present article and will be proven elsewhere.

The image under VAL will be decomposed into the following parts: Σ_0 – a part over the tip of the cone, which will be empty in our examples of even degree; Σ_∞ – a part at the infinite level, i.e. at the base of the cone, consisting of the intersection of a hypersurface and the quadric; Σ_R – the biggest part consisting of regular values, each of its slice at a given height will be consisting of the total space of the circle bundle $\mathcal{S}$ restricted to a symmetric bidegree curve on the quadric; Σ_C – the union of slices at critical levels each of which is an image of a section of $\mathcal{S}$ defined in the complement of a union of the two curves to the sides of the critical level, together with the whole fiber at their intersection.

A (phase) tropicalization can be defined as the closure of an image under (phase) valuation of a variety – note that the images under the valuation and coamoeba map generally are not closed, in contrast with usual (Archimedean) amoebas which are always closed.

We believe that the examples described below are actually typical, i.e. a very specific choice of the form of a polynomial is not crucial – allowing perhaps, obvious degeneration of critical levels. Moreover, analogously to [34] we, expect that in the generic case the corresponding closure is a topological manifold homeomorphic to a smooth complex surface of the corresponding degree.

Here, on the other hand, we emphasize that we are not going to take the closure, keeping the gaps at the attachment of regular and critical levels. This will be crucial in observing no lines in the next section.

Theorem 5.1. *Let n be a natural number and S_{2n} be a surface in $\mathbb{K}P^3$ of degree $2n$ given by homogeneous equation $F_{2n} = 0$, where*

$$F_{2n}(A) = \sum_{j=0}^{n} t^{-j(j+1)}(\det(A))^{n-j} f_{2j}(A),$$

where f_{2j} is a degree $2j$ homogeneous polynomial with complex coefficients not divisible by det *considered as a degree 2 polynomial. Denote by C_{2j} a complex curve on $Q(\mathbb{C})$ defined by the restriction of f_{2j}. Then*

$$\mathrm{VAL}(S_{2n}) \subset \Sigma_\infty \cup \Sigma_R \cup \Sigma_C,$$

where $\Sigma_\infty = \{\infty\} \times C_{2n}$, $\Sigma_R = \cup_{j=1}^{n-1}(j, j+1) \times \mathcal{S}|_{C_{2j}} \cup (n, \infty) \times \mathcal{S}|_{C_{2n}}$ and $\Sigma_C = \cup_{j=1}^{n}\{j\} \times (\sigma_j(Q(\mathbb{C})\backslash(C_{2j-2} \cup C_{2j})) \cup \mathcal{S}|_{C_{2j-2}\cap C_{2j}})$ and $\sigma_j[B]_{\mathbb{C}^*} = [\sqrt{-\frac{f_{2j-2}(B)}{f_{2j}(B)}}B]_{\mathbb{R}^*}$ is a section of $\mathcal{S}|_{Q(\mathbb{C})\backslash(C_{2j-2}\cup C_{2j})}$.

Proof. Take $A = t^\beta B + o(t^\beta)$ such that $F_{2n}(A) = 0$. Assume first $\det(A) = 0$. This immediately implies $f_{2n}(A) = 0$. From the last two equations we get two conditions on $B : \det(B) = 0$ and $f_{2n}(B) = 0$. Therefore, $\mathrm{VAL}(A) \in \Sigma_\infty$.

If A is invertible, normalize it to have $\det(A) = 1$. Observe again that this condition implies $\beta \geq 0$. However, if $\beta < 1$, the first term $\det(A)^n f_0(A)$ will dominate all others. This term, is, however, nonzero, therefore such a situation is not possible. In particular, $\mathrm{VAL}(S_{2n})$ has no points at heights in $[0, 1)$.

More generally, substitution in $F_{2n}(A)$ of all determinants with 1, results in an inhomogeneous polynomial that we further expand as a series in t. The expected valuation of a top order term in $F_{2n}(A)$ is given by the tropical polynomial $\max_{j=0}^{n}(-j(j+1) + 2j\beta)$, which has roots at $1, \dots, n$. If β is not one of these roots, $F_{2n}(A)$ has exactly one top order term $t^{-j(j+1)+2j\beta} f_{2j}(B)$ which must vanish. This implies that $[B]_{\mathbb{C}^*}$ belongs to C_m and $\mathrm{VAL}(A) \in \Sigma_R$.

In the situation when $j = \beta \in \{1, \dots, n\}$, there are two competing terms in $F_{2n}(A)$, the sum of which must vanish. We get an inhomogeneous equation $f_{2j-2}(B) + f_{2j}(B) = 0$ on B. If both $f_{2j-2}(B)$ and $f_{2j}(B)$ do not vanish (i.e. if $[B]_{\mathbb{C}^*} \in Q(\mathbb{C})\backslash(C_{2j-2} \cup C_{2j})$), this equation has a solution $cB \in [B]_{\mathbb{C}^*}$ for some $c \in \mathbb{C}^*$. Explicitly, by substituting cB into the inhomogeneous equation we get $c = \sqrt{-\frac{f_{2j-2}(B)}{f_{2j}(B)}}$ and $[B]_{\mathbb{C}^*} \mapsto [cB]_{\mathbb{R}^*}$ defines the section σ_j in the statement. That gives a point of Σ_C.

Finally, the case when one of $f_{2j-2}(B)$ or $f_{2j}(B)$ vanishes but the other does not, contradicts the assumption $F_{2n}(A) = 0$. If both vanish, $[B]_{\mathbb{C}^*} \in C_{2j} \cup C_{2j-2}$ and we do not put any additional restriction on B, landing under VAL to the fiber of $\mathcal{S}$ over $[B]_{\mathbb{C}^*}$ □

Theorem 5.2. *Let n be a natural number and S_{2n+1} be a surface in $\mathbb{K}P^3$ of degree $2n+1$ given by homogeneous equation $F_{2n+1} = 0$, where*

$$F_{2n+1}(A) = \sum_{j=0}^{n} t^{-j(j+1)}(\det(A))^{n-j} f_{2j+1}(A),$$

where f_{2j+1} is a degree $2j+1$ homogeneous polynomial with complex coefficients not divisible by det *considered as a degree 2 polynomial. Denote by C_{2j+1} a complex curve on $Q(\mathbb{C})$ defined by the restriction of f_{2j+1}. Then*

$$\mathrm{VAL}(S_{2n+1}) \subset \Sigma_\infty \cup \Sigma_R \cup \Sigma_C \cup \Sigma_0,$$

where $\Sigma_\infty = \{\infty\} \times C_{2n+1}$, $\Sigma_R = \cup_{j=0}^{n-1}(j, j+1) \times \mathcal{S}|_{C_{2j+1}} \cup (n, \infty) \times \mathcal{S}|_{C_{2n+1}}$ and $\Sigma_C = \cup_{j=1}^{n}\{j\} \times (\sigma_j(Q(\mathbb{C})\backslash(C_{2j-1} \cup C_{2j+1})) \cup \mathcal{S}|_{C_{2j-1} \cap C_{2j+1}})$ and $\sigma_j[B]_{\mathbb{C}^*} = [\sqrt{-\frac{f_{2j-1}(B)}{f_{2j+1}(B)}}B]_{\mathbb{R}^*}$ is a section of $\mathcal{S}|_{Q(\mathbb{C})\backslash(C_{2j-1}\cup C_{2j+1})}$, and $\Sigma_0 = \{0\} \times \varkappa^\circ(H)$, where H is the intersection with $PSL_2(\mathbb{C})$ of a plane in $\mathbb{C}P^3$ defined by $f_1 = 0$.

Proof. The argument is quite analogous to the even degree case, except for the situation when $\beta = 0$ in the asymptotic expansion $A = Bt^\beta + o(t^\beta)$. Under the assumption $\det(A) = 1$ this implies $\det(B) = 1$ and by substituting such A to F_{2n+1}, its leading term is $f_1(B)(\det(A))^n = f_1(B)$ the vanishing of which defines H, an intersection of a plane with $PSL_2(\mathbb{C})$. Clearly, $\mathrm{VAL}(A) \in \Sigma_0$. □

We expect that the inclusions in Theorems 5.1 and 5.2 are actually equalities. Moreover, allowing critical levels to vary and be multiple (in which case the sections would be multi-sections), as well as permitting for higher degree coamoebas above the tip of the cone in both even and odd cases, the description of the VAL images of surfaces is likely to be complete.

6. Lines on surfaces

In what follows, we combine the results of the previous two sections to illustrate how accurate is PSL_2 phase tropicalization in describing the relation between generic surfaces and lines in $\mathbb{P}^3$. As it was explained, the ordinary phase-forgetting PSL_2 tropicalization is hopeless in capturing this relation, but recovering the phase helps to restore the most characteristic elements of the picture.

First, observe that $\mathrm{VAL}(S_1)$ admits a two-dimensional family of VAL-images of lines. If f_1 is generic, then C_1 is an irreducible $(1,1)$-curve, in which case only types (2) and (4) can fit to $\mathrm{VAL}(S_1)$, since (1) and (3) have a section over either $(1,0)$-curve or $(0,1)$-curve at a non-zero level. The first type is parametrized by a single (double) point on C_1 and the second is parametrized by two distinct points on the same curve. Second, $\mathrm{VAL}(S_2)$

has one critical level at height 1, being empty for heights less than one, which implies that types (2) and (4) cannot belong to it. After height 1, it is spanned by the total space of $\mathcal{S}$ over the curve C_2. Assuming C_2 is irreducible, prohibits the types 1 and 3 if $r > 1$. Thus, r can be only equal to 1 and we get a one-dimensional family of lines consisting of two disjoint components.

Third, $\mathrm{VAL}(S_3)$ also has a single critical level. We will show now that every type of a VAL image of a line is rigid on $\mathrm{VAL}(S_3)$ provided that f_1 and f_3 are generic. The types (2) and (4) are rigid since they cannot be pushed from the tip of the cone by the irreducibility of C_1 and their rays must be based at the finite intersection of C_1 and C_3. For the types 1 and 3 again r must be equal to 1. Moreover, at each intersection of the line l with $C_3 \backslash C_1$ there must be a gap (where the section σ_1 is not defined) where a ray is attached. Note that in general, there are three such intersections of l and C_3, but we may have at most two rays on the image of VAL for a line. Thus, the only possibility is that l passes through one of the 6 intersection points of C_1 and C_3, which together with the choice of the bidegree (either $(0,1)$ or $(1,0)$) completely determines l.

As for higher degrees of surfaces, we have the following.

Theorem 6.1. *Let d be a natural number greater than 3 and let S_d be a degree d surface defined in either in Theorem 5.1 or 5.2 such that all the polynomials f_j are generic. Then, there exists no line L on $\mathbb{K}P^3$ such that $\mathrm{VAL}(L) \subset \mathrm{VAL}(S_d)$.*

Proof. Let's start with types (2) and (4). Note that $\mathrm{VAL}(S_d)$ has at least two critical levels, and the rays in the images of lines would need to penetrate through them, passing through the intersection of three generic curves on $Q(\mathbb{C})$, which is empty.

As for other types, their critical level r must be at the critical level of $\mathrm{VAL}(S_d)$ since all the curves C_j are irreducible by the genericity assumption. By the previous argument, r can be only n or $n-1$, where n is the rounding down of $\frac{d}{2}$, because the rays should penetrate through all the corresponding regular levels.

Assume first $r = n$. Then, l must have at most two intersection points with $(C_d \cup C_{d-2}) \backslash (C_d \cap C_{d-2})$ where the rays could be attached. However, by genericity of C_d its projection to one of the factors of $Q(\mathbb{C}) \cong \mathbb{C}P^1 \times \mathbb{C}P^1$ has only a simple ramification profile of order two, with ramification points having different images. Moreover, by genericity of C_{d-2}, this ramification points are distinct from the images of the intersection of C_{d-2} with C_d. Thus, a line l may have an intersection point of multiplicity two with C_d, but then all other intersections are transverse (with their total number equal $d-2$) and are all away from C_{d-2}, giving at least 3 intersections of l with $(C_d \cup C_{d-2}) \backslash (C_d \cap C_{d-2})$.

Finally, assume $r = n-1$. Then, the rays must penetrate through two regular levels and their endpoints must be at the intersection of two curves C_d and C_{d-2}. Moreover, neither of these intersections belongs to the same line l by genericity. Thus $\mathrm{VAL}(L)$, if it belongs to $\mathrm{VAL}(S_d)$ must have a single ray, and l must pass through its endpoint. On the other hand, again by genericity, l will not pass through $C_{d-2} \cap C_{d-4}$, so it will have $d-2-1 \geq 1$ additional gaps since all the intersections of l with C_{d-2} are transverse. A contradiction. □

Here we once more focus attention on the importance of not taking the closure of VAL(S_d), since otherwise we would lose the gap argument we used heavily above. Observe that in the geometric version of tropicalization, where we would take the Hausdorff (or Kuratowski) limit of a family, the result would be automatically close. It is a curious problem if the gaps could be restored in this setup.

7. Perspectives

We emphasize that our point is not reproving the well known fact of no-lines on generic degree≥ 4 surfaces, but rather showing the consistency of PSL_2 phase tropical geometry with the classical complex geometry at this basic level, something which is not there if the phase would be forgotten (the difference is less dramatic in the case of the usual tropicalization). Starting in that way there is a possibility for further exploration and adjustment. Notably, it might be helpful to further extend our tropicalization procedure to solve more elaborate questions, such as enumerative problems.

One such possibility for further refinement would be redefining VAL as $t^\alpha B + o(t^\alpha) \to (\alpha, [B]_{\{\pm 1\}})$ which would mean that instead of circle bundle $\mathcal{S}$ we would be dealing with a $\mathbb{C}^*$-bundle in a way more natural for complex geometry. It also might happen that for particular tasks forgetting the bundle altogether is enough, which is certainly true for the purposes of basic intersection theory. Such potential adjustments are aligned with the question posed by Oleg Viro. Namely, what kind of hyperfield-type structure on matrices produces shapes appearing in Theorems 5.1 and 5.2?

Of course, there is still much that remains unexplored. Remarks and suggestions would be much welcome. As was mentioned, we expect that the inclusions of Theorems 5.1 and 5.2 above are actually equalities.

References

[1] G. Mikhalkin and J. Rau, *Tropical geometry,* in preparation, available at: `https://math.uniandes.edu.co/~j.rau/downloads/main.pdf`.

[2] D. Maclagan and B. Sturmfels, *Introduction to tropical geometry,* American Mathematical Society, **161**, (2021).

[3] G. Mikhalkin and M. Shkolnikov, *Non-commutative amoebas,* Bulletin of the London Mathematical Society, **54(2)**, (2022), 335–368.

[4] M. Shkolnikov and P. Petrov, *Introduction to PSL_2 phase tropicalization,* in Proceedings of the Bulgarian Academy of Sciences, **77**, (2024), 1425–1432.

[5] M. Ruggiero and K. Shaw, *Tropical Hopf manifolds and contracting germs,* Manuscripta Mathematica, **152**, (2017), 1–60.

[6] G. Mikhalkin, *Examples of tropical-to-lagrangian correspondence,* European Journal of Mathematics, **5(3)**, (2019), 1033–1066.

[7] D. Matessi, *Lagrangian submanifolds from tropical hypersurfaces,* International Journal of Mathematics, **32(07)**, (2021), 2150046.

[8] M. Gross, *Tropical geometry and mirror symmetry,* American Mathematical Society, **(114)**, (2011).

[9] E. Brugallé, I. Itenberg, G. Mikhalkin, and K. Shaw, *Brief introduction to tropical geometry,* in Proceedings of the 21st Gökova Geometry-Topology Conference, (2015), 1–75.

[10] M. Krasner, *Approximation des corps valués complets de caractéristique $p \neq 0$ par ceux de caractéristique* 0, Colloque d'algèbre supérieure, tenu à Bruxelles du 19 au 22 décembre, (1956), 129–206.

[11] O. Viro, *Hyperfields for tropical geometry I. Hyperfields and dequantization,* arXiv:1006.3034, (2010).
[12] J. Giansiracusa and S. Mereta, *A general framework for tropical differential equations,* Manuscripta Mathematica, **173(3)**, (2024), 1273–1304.
[13] M. Joswig and B. Schröter, *Parametric shortest-path algorithms via tropical geometry,* Mathematics of Operations Research, **47(3)**, (2022), 2065–2081.
[14] L. Pachter and B. Sturmfels, *Algebraic statistics for computational biology,* Cambridge University Press, **13**, (2005).
[15] G. L. Litvinov and V. P. Maslov, *Idempotent mathematics: a correspondence principle and its applications to computing,* Russian Mathematical Surveys, **51(6)**, (1996).
[16] V. N. Kolokoltsov and V. P. Maslov, *Idempotent analysis and its applications,* vol. 401. Springer Science & Business Media, (2013).
[17] A. N. Gorban and O. Radulescu, *Dynamic and static limitation in multiscale reaction networks, revisited,* Advances in Chemical Engineering, **34**, (2008), 103–173.
[18] S. S. Samal, A. Naldi, D. Grigoriev, A. Weber, N. Théret, and O. Radulescu, *Geometric analysis of pathways dynamics: application to versatility of TGF-β receptors,* Biosystems, **149**, (2016), 3–14.
[19] L. Zhang, G. Naitzat, and L.-H. Lim, *Tropical geometry of deep neural networks,* in International Conference on Machine Learning, PMLR, (2018), 5824–5832.
[20] N. Grigg and N. Manwaring, *An elementary proof of the fundamental theorem of tropical algebra,* arXiv:0707.2591, (2007).
[21] H. Hoffman, *Pascal's mystic hexagon in tropical geometry,* HMC Senior Theses. 236, (2020).
[22] G. Rimmasch, *Complete tropical Bézout's theorem and intersection theory in the tropical projective plane,* PhD thesis, (2008).
[23] M. Panizzut and M. D. Vigeland, *Tropical lines on cubic surfaces,* SIAM Journal on Discrete Mathematics, **36(1)**, (2022), 383–410.
[24] G. Mikhalkin, *Enumerative tropical algebraic geometry in* $\mathbb{R}^2$, Journal of the American Mathematical Society, **18(2)**, (2005), 313–377.
[25] L. Caporaso and J. Harris, *Counting plane curves of any genus,* Inventiones Mathematicae, **131**, (1998), 345–392.
[26] I. Itenberg, V. Kharlamov, and E. Shustin, *Logarithmic equivalence of Welschinger invariants and Gromov-Witten invariants,* arXiv:math.AG/0407188, (2004).
[27] A. Gathmann and H. Markwig, *Kontsevich's formula and the WDVV equations in tropical geometry,* Advances in Mathematics, **217(2)**, (2008), 537–560.
[28] B. Sturmfels and J. Yu, *Tropical implicitization and mixed fiber polytopes,* Software for algebraic geometry, (2008), 111–131.
[29] E. Brugallé and K. Shaw, *Obstructions to approximating tropical curves in surfaces via intersection theory,* Canadian Journal of Mathematics, **67(3)**, (2015), 527–572.
[30] I. M. Gelfand, M. M. Kapranov, and A. V. Zelevinsky, *Discriminants, Resultants, and Multidimensional Determinants,* Birkhäuser Boston, (1994).
[31] G. Mikhalkin, *Amoebas of algebraic varieties and tropical geometry,* in Different faces of geometry, Springer, (2004), 257–300.
[32] O. Amini, M. Baker, and X. Faber, *Tropical and non-Archimedean geometry,* American Mathematical Society, **605**, (2014).
[33] W. P. Thurston, *Three-Dimensional Geometry and Topology, Volume 1,* Princeton University Press, (1997).
[34] G. Kerr and I. Zharkov, *Phase tropical hypersurfaces,* Geometry & Topology, **22(6)**, (2018), 3287–3320.

Institute of Mathematics and Informatics at the Bulgarian Academy of Sciences
Akad. G. Bonchev St, Bl. 8, 1113 Sofia, Bulgaria
Email address: m.shkolnikov@math.bas.bg

Institute of Mathematics and Informatics at the Bulgarian Academy of Sciences
Akad. G. Bonchev St, Bl. 8, 1113 Sofia, Bulgaria
Email address: pk5rov@gmail.com

Proceedings of 29th Gökova
Geometry-Topology Conference
pp. 18 – 36

Descent with algebraic structures for symplectic cohomology

Umut Varolgunes

ABSTRACT. We formulate and prove a chain-level descent property of symplectic cohomology for involutive covers by compact subsets that take into account the natural algebraic structures that are present. The notion of an involutive cover is reviewed. We indicate the role that the statement plays in mirror symmetry.

1. Introduction

Let (M^{2n}, ω) be a geometrically bounded symplectic manifold [1] and let $\mathbb{k}$ be a field of characteristic 0. If $c_1(TM) = 0$, we also fix a grading of M, that is, a homotopy class of non-vanishing sections of the complex line bundle $\Lambda^n_{\mathbb{C}}(TM) \to M$ for some compatible almost complex structure, to equip all the chain complexes below with $\mathbb{Z}$-gradings[1]; otherwise we only have $\mathbb{Z}_2$-gradings. The $\mathbb{k}$-algebra

$$\Lambda_{\geq 0} := \{\sum_{i\geq 0} a_i T^{\alpha_i} \mid a_i \in \mathbb{k}, \alpha_i \in \mathbb{R}_{\geq 0}, \text{ where } \alpha_i \to \infty, \text{ as } i \to \infty\}$$

is called the Novikov ring and its quotient field Λ is called the Novikov field.

The relative symplectic cohomology $SH^*_M(K)$ of a compact subset $K \subset M$ was defined in [2] using the geometric ideas of [3, 4] (see [1, 5, 6, 7] for similar constructions). Due to the reasoning explained in Remark 3.1, in this note we will try out a new terminology where we replace "relative symplectic cohomology" with "symplectic cohomology with supports" and "relative symplectic cohomology of K inside M" with "symplectic cohomology of M with support on K".

$SH^*_M(K)$ is the homology of a canonically defined chain complex $SC^*_M(K)$ over $\Lambda_{\geq 0}$ [8, Section 1.5], which we now call symplectic cochains of M with support on K. $SC^*_M(K)$ is obtained as the completed homotopy colimit of the universal homotopy coherent diagram of Floer complexes $CF^*(-;\Lambda_{\geq 0})$ of non-degenerate Hamiltonians that are negative on K with monotone continuation maps [2]. It can intuitively be thought of as the Floer complex of the upper semi-continuous function that is 0 on K and $+\infty$ outside.

---

[1]We use chain complex to mean cochain complex throughout the paper. This means that in our conventions, the differential of a chain complex increases the degree by one.

[2]Using $SC^*_M(K)$ for this canonical model is a change in notation as well. Previously, $SC^*_M(K)$ was used to denote the telescope model from [2], which was not ideal notation since it hid the choice of acceleration data involved.

There are canonical restriction chain maps $SC^*_M(K') \to SC^*_M(K)$, for $K \subset K'$, which upgrades the structure to a presheaf of chain complexes over the compact subsets of M (c.f. [8, Remark 1.6]). We also have a canonical PSS chain map [9]: $C^*(M;\mathbb{Z}) \otimes \Lambda_{>0} \to SC^*_M(K)$. Crucially, this map is a quasi-isomorphism if M is closed and $M = K$ [10, Section 3.3.3].

We cannot hope to have a local-to-global property for symplectic cohomology with supports in general because of the PSS isomorphism that we just mentioned and the fact that $SH^*_M(K) \otimes_{\Lambda_{\geq 0}} \Lambda$ is known to vanish for displaceable sets [11, 10], e.g. for sufficiently small Darboux balls. It turns out that for a special class of covers we have a satisfactory positive result.

Definition 1. *We say that the compact subsets $K_1, \dots, K_N \subset M$ are Poisson commuting if there exists a smooth map $F : M \to \mathbb{R}^k$ with Poisson commuting components and compact $P_1, \dots, P_N \subset \mathbb{R}^k$ such that $K_m = F^{-1}(P_m)$ for all $m = 1, \dots, N$. If $K = K_1 \cup \dots \cup K_N$ and $K_1, \dots, K_N$ are Poisson commuting compact subsets, we call $K_1, \dots, K_N$ an involutive cover of K.*

The Čech chain complex of $SC^*_M(-)$ for a cover $K = K_1 \cup \dots \cup K_N$ is defined as

$$\check{\mathrm{C}}\mathrm{ech}(SC_M; K_1, \dots, K_N) := \bigoplus_{p\geq 0} \bigoplus_{|J|=p+1} SC^*_M(\bigcap_{m\in J} K_m)[p],$$

where $[-]$ denotes a degree shift and J runs through subsets of $\{1, \dots, N\}$, and with differential the sum of the Čech differential and the Floer differential - ignoring the signs for the moment.

Theorem 1.1 ([2, Theorem 4.8.5]). *Assume that $K = K_1 \cup \dots \cup K_N$ is an involutive cover. Then, the canonical chain map*

$$SC^*_M(K) \to \check{\mathrm{C}}\mathrm{ech}(SC_M; K_1, \dots, K_N) \tag{1}$$

is a quasi-isomorphism.

In fact, below we give a very mild generalization of this result to weakly involutive covers (Definition 2) in Theorem 3.5, which in particular gives us a chance to review the proof.

In [8], with Yoel Groman and Mohammed Abouzaid, we construct a bigger model $SC^*_{M,big}(K)$ of $SC^*_M(K)$, which has a natural action of the Kimura-Stasheff-Voronov (KSV) operad (see Section 2.2 for our specific usage here and references) and restriction maps that are compatible with this action upgrading the structure to a presheaf of KSV operad algebras (again, see [8, Remark 1.6]). Here, a bigger model precisely means that we have canonical inclusions of chain complexes

$$SC^*_M(K) \to SC^*_{M,big}(K),$$

compatible with restrictions maps, that induce isomorphisms on homology.

The goal of this note is to incorporate this algebraic structure into the local-to-global principle for weakly involutive covers. We achieve this task in Theorem 4.1 using existing techniques from the literature. We then indicate how this result would be used in mirror symmetry in the final section.

2. Background

2.1. Weakly involutive covers

Let M be a symplectic manifold, which has an induced Poisson bracket $\{-,-\} : C^\infty(M) \times C^\infty(M) \to C^\infty(M)$. We note the following well-known result, e.g. see [2, Lemma 4.8.4].

Lemma 2.1. *Let $f_1, \ldots, f_N : M \to \mathbb{R}$, and $g_1, g_2 : \mathbb{R}^N \to \mathbb{R}$ be smooth functions. Assume that $\{f_i, f_j\} = 0$, for all i, j. Then the functions $G_l : M \to \mathbb{R}$, $l = 1, 2$, defined by $G_l(x) = g_l(f_1(x), \ldots, f_N(x))$ also satisfy $\{G_1, G_2\} = 0$.* □

Definition 2. *We call compact subsets $K_1, ..., K_N \subset M$ weakly Poisson commuting if there exists smooth functions $f_{m,i} : M \to \mathbb{R}$ for $m = 1, \ldots, N$ and $i = 1, 2, \ldots$ such that*

- *$f_{m,i} |_{K_m} < 0$ for all m and i.*
- *$f_{m,i} < f_{m,i+1}$ for all m and i.*
- *For all m, if $f_{m,i}(x) < 0$ for all i, then $x \in K_m$.*
- *The Poisson bracket $\{f_{m,i}, f_{m',i}\} = 0$ for all i and m, m'.*

If $K = K_1 \cup \ldots \cup K_N$ and $K_1, \ldots, K_N$ are weakly Poisson commuting compact subsets, we call $K_1, \ldots, K_N$ a weakly involutive cover of K.

The order of the compact subsets is, of course, unimportant. We used this notation for ease of reading. The weakly Poisson commutation property is inherited by sublists.

Proposition 2.2. *If $K_1, ..., K_N$ are Poisson commuting as in Definition 1, then they are weakly Poisson commuting.*

Proof. This immediately follows from the fact that for compact $Z \subset \mathbb{R}^k$, we can always find a non-negative smooth function on $\mathbb{R}^k$ which vanishes precisely on Z along with Lemma 2.1. □

Remark 2.3. *We do not know whether weakly Poisson commuting is a strictly weaker condition than Poisson commuting, but we expect it to be more convenient in the "degeneration setup". An example of this setup is a semi-stable polarized degeneration of a smooth complex manifold over a disk* [12, 13]. *Roughly speaking, we use symplectic parallel transport to move the decomposition of the central fiber into its irreducible components to the general fiber in order to obtain an involutive cover using a version of* [14, Theorem 2.12] *(also see the more recent* [15]*).*

Proposition 2.4. *Let K be a finite union of finite intersections of the compact subsets $K_1, \ldots, K_N$ for which there exists smooth functions $f_{m,i} : M \to \mathbb{R}$ with $m = 1, \ldots, N$ and $i = 1, 2, \ldots$ such that*

- *$f_{m,i} |_{K_m} < 0$ for all m and i.*

- $f_{m,i} < f_{m,i+1}$ *for all* m *and* i.
- *For all* m, *if* $f_{m,i}(x) < 0$ *for all* i, *then* $x \in K_m$.

Then, we can construct smooth functions $g_i : M \to \mathbb{R}$, $i = 1, 2, \dots$ *such that*

- $g_i \mid_K < 0$ *for all* i.
- $g_i < g_{i+1}$ *for all* i.
- *If* $g_i(x) < 0$ *for all* i, *then* $x \in K$.
- *There are smooth* $h_i : \mathbb{R}^N \to \mathbb{R}$ *such that*

$$g_i(x) = h_i(f_{1,i}(x), \dots, f_{N,i}(x))$$

for all i *and* x.

Proof. By a simple induction, it suffices to prove the case $N = 2$, where K is either $K_1 \cap K_2$ or $K_1 \cup K_2$.

We deal with the intersection case. Consider the maps $F_i := (f_{1,i}, f_{2,i}) : M \to \mathbb{R}^2$. By our assumption, F_i's map K_1 to the open left half space and K_2 to the open lower half space. Let C_i be a curve that is a smoothing of the union of the non-positive parts of the two axes in $\mathbb{R}^2$, which is contained in the negative quadrant. We assume that the slopes of all of its tangent lines are non-positive and no negative slope is attained twice. The complement of C_i in the plane has two components. We call the one that is contained in the negative quadrant B_i. We can make sure that $F_i(K_1 \cap K_2) \subset B_i$ and $B_{i+1} \supset B_i$ for all i. For example, we can take

$$C_i = \{xy = \delta_i \mid x, y < 0\},$$

with δ_i a sequence of positive numbers monotonically converging to 0 sufficiently fast.

We define a smooth function $h_i : \mathbb{R}^2 \to \mathbb{R}$ which is 0 on C_i, negative on B_i and positive elsewhere as follows: Let $x \in \mathbb{R}^2$, draw the straight line with slope 1 from x and define $h_i(x)$ to be the signed length (negative on B_i and positive on the other component) of the segment between x and where it intersects C_i. It is now elementary to check that $g_i := h_i \circ F_i$ satisfies the conditions.

The union case is similar and we omit it. □

Corollary 2.5. *If* $K_1, ..., K_N$ *are weakly Poisson commuting, then the list of all compact subsets that can be obtained as a finite union of finite intersections of* $K_1, \dots, K_N$ *is also weakly Poisson commuting.*

Proof. This is immediate from Proposition 2.4 and Lemma 2.1. □

Note that this corollary is true when we remove the words "weakly" as well, and the proof is easier.

2.2. The KSV operad

We will use the notions of operads and algebras over operads freely. For definitions, see [16, Section I.1-2]. Unless otherwise specified, our operations are symmetric and over

$Ch(\Lambda_{\geq 0})$. We also do not consider 0-ary operations (e.g. units of algebras) throughout the paper.

The operad that is relevant here is the KSV operad [17, 8]

$$\{\mathrm{KSV}(n) = C_{-*}(f\overline{\mathcal{M}}^{\mathbb{R}}_{0,n+1}; \Lambda_{\geq 0})\}_{n\geq 1}.$$

Here $f\overline{\mathcal{M}}^{\mathbb{R}}_{0,2} := S^1$ and, for $n \geq 2$, an element of $f\overline{\mathcal{M}}^{\mathbb{R}}_{0,n+1}$ is a stable nodal curve from $\overline{\mathcal{M}}_{0,n+1}$ equipped with tangent rays at the branches of each node up to simultaneous rotation (in opposite directions) as well as at the marked points. Note that $f\overline{\mathcal{M}}^{\mathbb{R}}_{0,n+1}$ can be obtained by doing a real blow-up along the boundary divisor (a normal crossing divisor) of the Deligne-Mumford space $\overline{\mathcal{M}}_{0,n+1}$. Finally, C_{-*} denotes a version of cubical singular chains, symmetric normalized cubical singular chains to be specific, with negated grading. The details are unimportant for the purposes of this paper.

It is well-known that algebras over the homology operad of KSV are precisely BV algebras [18]. The definition of the latter is reviewed in Section 3.3. In other words, the homology operad of KSV is isomorphic to the BV operad. Let us note that the KSV operad is formal [19, 20], see also [8, Proposition 1.11]. This formality is only true because we assumed that our ground field $\Bbbk$ is characteristic 0.

In [21], a cofibrant replacement for the BV operad (and by its formality, for the KSV operad) was constructed. Following the same paper, we call algebras over this operad homotopy BV algebras. For these, there is a more flexible notion of a morphism that we call BV_∞-maps [21, Section 4]. We have a functor from KSV-algebras with strict morphisms to homotopy BV algebras with BV_∞-maps.

2.3. Homotopy limits of cosimplicial diagrams

Let $\mathbb{N}^{inj}_{aug}$ be the category with objects $-1, 0, 1, 2, \dots$ and morphisms from p to q the set of injective order-preserving maps $\{n \in \mathbb{Z} \mid 1 \leq n \leq p+1\} \to \{n \in \mathbb{Z} \mid 1 \leq n \leq q+1\}$. $\mathbb{N}^{inj}$ denotes the full subcategory with non-negative integers as objects.

A contravariant/covariant functor $\mathbb{N}^{inj} \to Ch(\Lambda_{\geq 0})$ is called a semi-simplicial / semi-cosimplicial chain complex. An augmentation of a semi-cosimplicial chain complex $\mathbb{N}^{inj} \to Ch(\Lambda_{\geq 0})$ is simply an extension of the functor to $\mathbb{N}^{inj}_{aug} \to Ch(\Lambda_{\geq 0})$.

Let us denote by Δ^p the p dimensional simplex inside $\mathbb{R}^{p+1}$ and the affine, vertex order preserving inclusions into codimension one faces by $\delta_i : \Delta^p \to \Delta^{p+1}$, with $i = 0, \dots, p+1$.

We start with an alternative interpretation of $\check{\mathrm{C}}\mathrm{ech}(SC_M; K_1, \dots, K_N)$. Let $(\mathcal{D}^\bullet, d_i)$ be a semi-cosimplicial chain complex. In our case, this will be given by

$$\mathcal{SC}^\bullet := p \mapsto \bigoplus_{|J|=p+1} SC^*_M(\bigcap_{m\in J} K_m),$$

with face maps d_i obtained from the restriction maps in the standard way. An augmentation of $\mathcal{SC}^\bullet$ is given by

$$-1 \mapsto SC^*_M(K_1 \cup \dots \cup K_N).$$

Let us define the semi-simplicial chain complex

$$p \mapsto NC^*(\Delta^p) := \bigoplus_{F \subset \Delta^p} \Lambda_{\geq 0} \cdot \delta_F$$

where the direct sum is over all faces of Δ^p (which are in one to one correspondence with all the morphisms in $\mathbb{N}^{inj}$ with target n) and δ_F is in degree equal to the dimension of F. We think of the elements of $NC^*(\Delta^p)$ as assignments of scalars to each face of Δ^p. We equip $NC^*(\Delta^p)$ with the differential that sends δ_F to the alternating sum of $\delta_{F'}$, where F' is a face containing F of dimension 1 higher. The coface maps of $NC^*(\Delta^\bullet)$ are those induced by the face inclusions $\Delta^p \to \Delta^q$.

We can define the totalization of $\mathcal{D}^\bullet$ as the chain complex

$$Tot(\mathcal{D}^\bullet) := \mathrm{eq}\left(\prod_{p=0}^{\infty} D^p \otimes NC^*(\Delta^p) \rightrightarrows \prod_{f:r \to q} D^q \otimes NC^*(\Delta^r)\right).$$

Here the second product is over all maps in $\mathbb{N}^{inj}$ and equalizer means that we are taking the kernel of the difference of the two natural maps:

1. (1) whose $f : r \to q$ component is the projection $\prod_{p=0}^{\infty} D^p \otimes NC^*(\Delta^p) \to D^q \otimes NC^*(\Delta^q)$ composed with the coface map $f^* : NC^*(\Delta^q) \to NC^*(\Delta^r)$, and
2. (2) whose $f : r \to q$ component is the projection $\prod_{p=0}^{\infty} D^p \otimes NC^*(\Delta^p) \to D^r \otimes NC^*(\Delta^r)$ composed with the face map $f_* : D^q \to D^r$.

We find it helpful to think of the elements of $D^q \otimes NC^*(\Delta^r)$ as normalized simplicial cochains on Δ^r with values in D^q.

An augmentation of $\mathcal{D}^\bullet$ gives rise to a canonical map

$$D^{-1} \to Tot(\mathcal{D}^\bullet)$$

by sending $d \in D^{-1}$ to the element of $\prod_{p=0}^{\infty} D^p \otimes NC^*(\Delta^p)$ with p component equal to the degree 0 normalized simplicial cochain that sends every vertex of Δ^p to $f_*(d)$ for the unique arrow $f : 0 \to p$ in $\mathbb{N}^{inj}_{aug}$.

It is important to notice that the inclusion of $Tot(\mathcal{D}^\bullet)$ into

$$\{(x_p = (f \mapsto x_p(f))) \in \prod_{p=0}^{\infty} D^p \otimes NC^*(\Delta^p) \mid d_i(x_p(f)) = x_{p+1}(\delta_i f)$$
$$\text{for all } p \geq 0, i = 0, \ldots, p+1 \text{ and face } f \text{ of } \Delta^p\}$$

is surjective. The following is easy to see from this description.

Proposition 2.6. *There is an isomorphism of chain complexes*

$$Tot(\mathcal{SC}^\bullet) \to \check{\mathrm{C}}\mathrm{ech}(SC_M; K_1, \ldots, K_N)$$

by $(x_p) \mapsto (x_p(F))$, *where* F *denotes the unique codimension* 0 *faces. Moreover, the isomorphism intertwines the canonical maps received from* $SC^*_M(K)$, *where* $K = K_1 \cup \ldots \cup K_N$. □

In fact, $NC^*(\Delta^\bullet)$ is a semi-simplicial dga using the standard cup product. Therefore, if $\mathcal{D}^\bullet$ is a semi-cosimplicial algebra over a non-symmetric operad, $Tot(\mathcal{D}^\bullet)$ has an algebra structure over the same operad (from the same reasoning as in Proposition 2.9). This is commonly used in the literature for defining product structures in Čech complexes of dga's, e.g. [22, Equation (14.24)]. Unfortunately, the KSV operad is a symmetric operad. So we need a replacement for $NC^*(\Delta^\bullet)$ that has a semi-simplicial commutative dga structure (see Remark 2.10). Following Sullivan this is possible under the characteristic 0 assumption (see [23, 24, 25]).

Let us define the commutative dga (cdga) of polynomial differential forms on the p-simplex as

$$\Omega^*(\Delta^p) := \frac{\Lambda_{\geq 0}[t_0, \dots, t_p, dt_0, \dots, dt_p]}{(\sum t_i - 1, \sum dt_i)}$$

with $\deg(t_i) = 0$ and $\deg(dt_i) = 1$. Using pullback of forms one can easily define the semi-simplicial cdga:

$$p \mapsto \Omega^*(\Delta^p).$$

We then define the Thom-Whitney chain complex of $\mathcal{D}^\bullet$ by

$$TW(\mathcal{D}^\bullet) := \mathrm{eq}\left(\prod_{p=0}^{\infty} D^p \otimes \Omega^*(\Delta^p) \rightrightarrows \prod_{r \to q} D^q \otimes \Omega^*(\Delta^r)\right)$$

using the same notation with the definition of totalization. An augmentation gives rise to a canonical map

$$D^{-1} \to TW(\mathcal{D}^\bullet).$$

Proposition 2.7. *By integrating forms, we can define a map of semi-cosimplicial spaces:*

$$\Omega^*(\Delta^\bullet) \to NC^*(\Delta^\bullet),$$

which induces a quasi-isomorphism

$$TW(\mathcal{D}^\bullet) \to Tot(\mathcal{D}^\bullet)$$

compatible with the maps obtained from an augmentation.

Proof. One can even write down an explicit and universal homotopy equivalence between these two complexes extending the integration map, see [25, Section 7] and the references contained therein. □

Remark 2.8. *Using the projective model structure on unbounded chain complexes* [26, Theorem 2.3.11] *and the induced (injective) model structure on semi-simplicial complexes* [26, Theorem 5.1.3] *(using that $(\mathbb{N}^{inj})^{op}$ is an inverse category), one can show that Tot and TW both compute the homotopy limit of semi-cosimplicial chain complexes. This follows from* [27, Corollary 3.5]. *Noting that, by definition, a map of chain complexes is fibrant if it is surjective, the needed fibrancy in the injective model structure is obvious in the case of Tot and it follows from the basic* [28, Lemma 9.4] *for TW.*

Proposition 2.9. *Let $\mathcal{O}$ be a (symmetric) operad over $Ch(\Lambda_{\geq 0})$ and assume that we have a semi-cosimplicial object $\mathcal{A}^\bullet$ in $\mathcal{O}$-algebras. Then, $TW(\mathcal{A}^\bullet)$ is canonically an $\mathcal{O}$-algebra so that for an augmentation as $\mathcal{O}$-algebras, the canonical map $\mathcal{A}^{-1} \to TW(\mathcal{A}^\bullet)$ is a map of $\mathcal{O}$-algebras.*

Proof. The key point is that if C is a commutative dga, then $D \mapsto D \otimes C$ defines a symmetric lax monoidal functor $Ch(\Lambda_{\geq 0}) \to Ch(\Lambda_{\geq 0})$ with the natural transformation

$$(D_1 \otimes C) \otimes (D_2 \otimes C) \to (D_1 \otimes D_2) \otimes C$$

given by the product structure of C. This implies that if A is an $\mathcal{O}$-algebra, then so is $A \otimes C$ automatically. This means that $\prod_{p=0}^{\infty} A^p \otimes \Omega^*(\Delta^p)$ has a canonical $\mathcal{O}$-algebra structure.

Now, notice that the $\mathcal{O}$-algebra structure that we obtain (in the notation of the previous paragraph) on $A \otimes C$ is natural in both A and C. More precisely, if A_1, A_2 are $\mathcal{O}$-algebras, $A_1 \to A_2$ is a map of $\mathcal{O}$-algebras and C_1, C_2 are cdga's, $C_1 \to C_2$ is a map of algebras, then the natural map $A_1 \otimes C_1 \to A_2 \otimes C_2$ is a map of $\mathcal{O}$-algebras. Using this, it is easy to see that the $\mathcal{O}$-algebra structure on $\prod_{p=0}^{\infty} A^p \otimes \Omega^*(\Delta^p)$ descends to the equalizer defining $TW(\mathcal{A}^\bullet)$. □

Remark 2.10. *It might be worth pointing out at this point why we would not be able to use $NC^*(\Delta^\bullet)$ in the place of $\Omega^*(\Delta^\bullet)$ for the construction in the proof of Proposition 2.9. Instead of $\mathcal{O}$-algebras for a general symmetric operad $\mathcal{O}$, let us consider the example of commutative algebras first. The problem is that for A a commutative algebra, if the algebra structure on C is not commutative, then the algebra structure that is induced on $A \otimes C$ is also not going to be commutative. In the general $\mathcal{O}$-algebras case, the naive ansatz for defining an $\mathcal{O}$-algebra structure on $A \otimes C$ from the $\mathcal{O}$-algebra structure on A and the algebra structure on C will not satisfy the symmetric group equivariance axiom of an $\mathcal{O}$-algebra.*

Remark 2.11. *There is a model structure on $\mathcal{O}$-algebras transferred using the free-forgetful adjunction to chain complexes* [29, Theorem 4.1.1]*. This again induces the injective model structure on semi-simplicial $\mathcal{O}$-algebras. By definition, again, fibrancy of a map of $\mathcal{O}$-algebras means surjectivity. Therefore, by the same reasoning as Remark 2.8, TW computes the homotopy limit of semi-cosimplicial $\mathcal{O}$-algebras for this particular model structure.*

3. Symplectic cohomology with supports

3.1. Definition

Let us introduce symplectic cohomology with supports following [8, Section 1.5]. The details are not so important here and can be found in [2, 30, 8]. We hope to simply provide an overview to the reader. We also do not include the modifications that are necessary in the case where M is geometrically bounded but not closed and refer the reader to [8, Appendix C].

For a non-degenerate Hamiltonian $H : \mathbb{R}/\mathbb{Z} \times M \to \mathbb{R}$ we denote the Hamiltonian Floer cochain complex [31, 2] of H over $\Lambda_{\geq 0}$ by $CF^*(H;\Lambda_{\geq 0})$.

We first define a dg-category $\mathcal{H}^{cyl}$ whose objects are non-degenerate Hamiltonians. Morphisms from H_- to H_+ are (roughly speaking) chains on a carefully constructed cubical set of monotone continuation map data whose unbroken 0-cubes are

$$\{H_s : \mathbb{R} \times \mathbb{R}/\mathbb{Z} \times M \to \mathbb{R} \mid H_s = H_\pm \text{ near } s = \pm\infty \text{ and } \partial_s H_s \geq 0\}. \tag{2}$$

Composition is given by associating the outputs with inputs to form broken data.

We then construct the Floer functor

$$\mathcal{CF} : \mathcal{H}^{cyl} \to Ch^{dg}(\Lambda_{\geq 0}). \tag{3}$$

On objects it is defined by $H \mapsto CF^*(H;\Lambda_{\geq 0})$ and for morphisms we consider continuation maps defined by virtual dimension zero parametrized moduli spaces of Floer solutions. Moreover, $\mathcal{H}^{cyl}$ has a canonical functor to the dg-category $\star$ with one object and endomorphism space isomorphic to $\Lambda_{\geq 0}$. Note that the chain complex $\Lambda_{\geq 0}$ is not a terminal object in $Ch^{dg}(\Lambda_{\geq 0})$. This canonical augmentation exists because the category of cubical sets does have a terminal object whose normalized chains is equal to $\Lambda_{\geq 0}$.

Finally, for a compact $K \subset M$, $SC^*_M(K)$ is defined as the degreewise T-adic completion of the homotopy colimit of $\mathcal{CF}$ restricted to the full subcategory $\mathcal{H}^{cyl}_K$ with objects the non-degenerate Hamiltonians that are negative on K:

$$SC^*_M(K) := \widehat{\text{hocolim}}\left(\mathcal{CF}|_{\mathcal{H}^{cyl}_K}\right). \tag{4}$$

Homotopy colimit here is defined by a two sided bar complex $B(\star, \mathcal{H}^{cyl}_K, \mathcal{CF})$ [32, 33]. The underlying module of the bar complex is

$$\bigoplus_{n=0}^{\infty} \bigoplus_{(H_0,\ldots,H_n)\in Ob\mathcal{H}^{cyl}_K} CF^*(H_0;\Lambda_{\geq 0}) \otimes Hom_{\mathcal{H}^{cyl}}(H_0,H_1) \otimes \cdots \otimes Hom_{\mathcal{H}^{cyl}}(H_{n-1},H_n)$$

and the differential is obtained using composition in the category, the module action and the canonical augmentations of the morphism complexes.

The homology of $SC^*_M(K)$ is denoted by $SH^*_M(K)$ and is called the symplectic cohomology of M with support on K. It will be important for us to work at the chain complex level (keeping in mind for example the generalized Mayer-Vietoris principle for de Rham cohomology [22]).

Remark 3.1. *As already mentioned in the introduction, we had previously called $SH^*_M(K)$ the relative symplectic cohomology of K inside M. Let us explain the main reason for wanting to change terminology. The theory has an open-string counterpart, which is currently in development (see* [30, Section 2.3] *for baby steps in this direction). Eventually, there will be a Fukaya category that is associated to a compact subset. Without a change in terminology this would have to be called the relative Fukaya category which clashes with the Seidel-Sheridan terminology for the Fukaya category of a symplectic manifold relative to a divisor* [34, 35]*. We hope the new terminology catches on.*

There are canonical restriction chain maps

$$SC^*_M(K') \to SC^*_M(K), \tag{5}$$

for $K \subset K'$, which upgrades the structure to a presheaf over the compact subsets of M. The base change of the symplectic cochains with supports presheaf $SC^*_M(-)$ to the Novikov field will be denoted by $SC^*_M(-;\Lambda)$ and its homology by $SH^*_M(-;\Lambda)$.

For calculations, it is better to introduce a smaller model [2]. We define an acceleration datum for $K \subset M$ as a sequence $H_1 \leq H_2 \leq \dots$ of non-degenerate Hamiltonians $H_i : \mathbb{R}/\mathbb{Z} \times M \to \mathbb{R}$ satisfying $H|_{\mathbb{R}/\mathbb{Z}\times K} < 0$ and for every $(t,x) \in \mathbb{R}/\mathbb{Z} \times M$,

$$H_i(t,x) \xrightarrow[i\to\infty]{} \begin{cases} 0, & x \in K, \\ +\infty, & x \notin K. \end{cases}$$

along with monotone interpolations between H_i and H_{i+1} for all $i \geq 1$.

This pointwise convergence condition is equivalent to the subcategory of $\mathcal{H}^{cyl}_K$ with objects H_i and morphisms given by the span of the chosen 0-chains being homotopy cofinal. In the framework of [32, Proposition 4.4] (also see [8, Section 5.7]), this homotopy cofinality means that for any object G of $\mathcal{H}^{cyl}_K$ the canonical map from the homotopy colimit of

$$\mathcal{H}^{cyl}_K(G,H_1) \to \mathcal{H}^{cyl}_K(G,H_2) \to \mathcal{H}^{cyl}_K(G,H_3)\dots$$

to $\Lambda_{\geq 0}$ is a quasi-isomorphism (cf. the discussion of Equations (5.54)-(5.56) in [8]). It is elementary to show that acceleration data exist.

The homotopy cofinality implies that the canonical inclusion of the homotopy colimit of the diagram

$$\mathcal{C}_K := CF^*(H_1;\Lambda_{\geq 0}) \xrightarrow{\kappa_1} CF^*(H_2;\Lambda_{\geq 0}) \xrightarrow{\kappa_2} \dots$$

into $\operatorname{hocolim} \mathcal{CF}|_{\mathcal{H}^{cyl}_K}$ is a quasi-isomorphism (this is the canonical map in [8, Theorem 1.9]). It is well-known that this simpler homotopy colimit can also be computed by the telescope model

$$\mathrm{tel}(\mathcal{C}_K) := \mathrm{cone}\left(\bigoplus_{i=1}^{\infty} CF^*(H_i;\Lambda_{\geq 0}) \xrightarrow{\kappa - id} \bigoplus_{i=1}^{\infty} CF^*(H_i;\Lambda_{\geq 0})\right),$$

where κ is the linear map that sends the summand $CF^*(H_i;\Lambda_{\geq 0})$ to $\bigoplus_{i=1}^{\infty} CF^*(H_i;\Lambda_{\geq 0})$ by composing κ_i with the canonical inclusion. Denoting the degree-wise T-adic completion of $\mathrm{tel}(\mathcal{C}_K)$ by $\widehat{\mathrm{tel}}(\mathcal{C}_K)$ we, therefore, have a canonical chain map

$$\widehat{\mathrm{tel}}(\mathcal{C}_K) \to SC^*_M(K), \tag{6}$$

which is also a quasi-isomorphism by [2, Corollary 2.3.6(3)].

We note the following straightforward extension of [8, Theorem 1.8] without proof:

Proposition 3.2. *Assume that X, Y are compact subsets of M. Then, we have a homotopy commutative diagram:*

$$\begin{array}{ccc} \widehat{\mathrm{tel}}(\mathcal{C}_{X\cup Y}) & \longrightarrow & cocone\left(\widehat{\mathrm{tel}}(\mathcal{C}_X) \oplus \widehat{\mathrm{tel}}(\mathcal{C}_Y) \to \widehat{\mathrm{tel}}(\mathcal{C}_{X\cap Y})\right) \\ \downarrow & & \downarrow \\ SC^*_M(X\cup Y) & \longrightarrow & \check{\mathrm{C}}\mathrm{ech}(SC_M; X, Y) \end{array}$$

where the top row constructed using appropriate acceleration data as in [2, Section 3.4] *and the vertical arrows are quasi-isomorphisms.* □

Remark 3.3. *For a chain map $f : C \to D$ of $\mathbb{Z}_2$-graded chain complexes, the reference* [2] *uses the formula $C[1] \oplus D$, with the additional off-diagonal term in the differential given by f, for the definition of cone(f). In the $\mathbb{Z}$-graded context, the accurate formula is not $C[1] \oplus D$ but $C[-1] \oplus D$. Then, cocone(f) is simply the chain complex cone$(f)[1]$. It is important to notice that cone(f) receives a map from D and completes f to an exact triangle in that direction, whereas cocone(f) maps to C and completes f to an exact triangle in the opposite direction. Hence, following the standard usage of the co-terminology, it would have been better to say ne instead of cocone, but this is not standard.*

3.2. Descent at the linear level

We first state and prove the statement for two subsets.

Theorem 3.4. *Assume that $K = K_1 \cup K_2$ is a weakly involutive cover. Then, the canonical chain map*

$$SC^*_M(K) \to \check{\mathrm{C}}\mathrm{ech}(SC_M; K_1, K_2) \tag{7}$$

is a quasi-isomorphism.

Proof. Theorem 4.8.2 from [2] shows that we can make choices such that the upper horizontal arrow in the diagram of Proposition 3.2 is a quasi-isomorphism. Homotopy commutativity of the same diagram finishes the proof. □

We note that [2, Theorem 4.8.1] is more general in that it assumes Poisson commutativity only along the "barrier", but we do not know whether this has any use or not.

Theorem 3.5. *Assume that $K = K_1 \cup \ldots \cup K_N$ is a weakly involutive cover. Then, the canonical chain map*

$$SC^*_M(K) \to \check{\mathrm{C}}\mathrm{ech}(SC_M; K_1, \ldots, K_N) \tag{8}$$

is a quasi-isomorphism.

Before we give the proof, we give a lemma that is a purely algebraic statement about the presheaf of chain complexes $SC^*_M(-)$ paraphrasing [2, Appendix B]. If the conclusion of Theorem 3.5 holds, let us say that $K_1, \ldots, K_N$ satisfies descent.

Lemma 3.6. *Let $K_1, \ldots, K_N$ be compact subsets with $N > 2$. Assume that $K_1, K_2 \cup \ldots \cup K_N$ satisfies descent as well as $K_2, \ldots, K_N$ and $K_1 \cap K_2, \ldots, K_1 \cap K_N$. Then, $K_1, \ldots, K_N$ satisfies descent.*

Proof. We have that the canonical maps

$$SC^*_M(K_1 \cup \ldots \cup K_N) \to \check{\mathrm{C}}\mathrm{ech}(SC_M; K_1, K_2 \cup \ldots \cup K_N),$$

$$\check{\mathrm{C}}\mathrm{ech}(SC_M; K_1, K_2 \cup \ldots \cup K_N) \to cocone(SC^*_M(K_1) \oplus \check{\mathrm{C}}\mathrm{ech}(SC_M; K_2, \ldots, K_N) \\ \to \check{\mathrm{C}}\mathrm{ech}(SC_M; K_1 \cap K_2, \ldots, K_1 \cap K_N))$$

are quasi-isomorphisms. Noticing that the cocone is nothing but $\check{\mathrm{C}}\mathrm{ech}(SC_M; K_1, \ldots, K_N)$ completes the proof. □

Proof of Theorem 3.5. Let $\mathcal{K}$ be the smallest set of subsets of M that is closed under intersection and union, and contains $K_1, \ldots, K_n$. By Corollary 2.5, the members of $\mathcal{K}$ are Poisson commuting. By Theorem 3.4, for any two-element list from $\mathcal{K}$ descent is satisfied. Using Lemma 3.6, we finish the proof by induction. □

3.3. Algebraic structures

We now discuss algebraic structures in symplectic cohomology with supports. Let us start with the homology level structure.

A BV-algebra is a graded commutative algebra A equipped with a degree decreasing differential Δ, called the BV operator, with the following property. It does not satisfy the graded Leibniz rule, but the error can be used to define a degree -1 Lie bracket on A, which, in particular, does satisfy the graded Leibniz rule in both of its slots. We recommend [36, Section 2.2] for the precise definition. A unit in a BV-algebra is a multiplicative unit that is Δ-closed.

Using the techniques of [30], $SH^*_M(K)$ can be equipped with a natural BV-algebra structure, which also admits a unit after base change to Λ. The restriction maps respect these structures.

We now move on to the chain-level structure, the statement of which requires the terminology from Section 2.2.

Theorem 3.7 ([8]). *Associated to any compact subset $K \subset M$ is a complete torsion-free chain complex $SC^*_{M,big}(K)$ over the Novikov ring, which is equipped with the following structures:*

1. *An action of the KSV operad.*
2. *A restriction map for each inclusion $K \subset K'$ of compact subsets*

$$SC^*_{M,big}(K') \to SC^*_{M,big}(K),$$

which is compatible with the operadic action and satisfies the properties of a presheaf.

(3) *There is a canonical quasi-isomorphism*

$$SC^*_M(K) \to SC^*_{M,big}(K),$$

which is compatible with restriction maps.

Proof. See Theorem 1.4 and Theorem 1.9 of [8]. □

The BV algebra structure that we obtain on $H^*(SC^*_{M,big}(K)) \simeq SH^*_M(K)$ agrees with the aforementioned one. We also obtain a canonical homotopy BV-algebra structure on $SC^*_{M,big}(K)$.

Remark 3.8. *In fact, this is not the full algebraic structure of symplectic cohomology with support, as one can consider the operations induced by positive genus Riemann surfaces and also allow more than one output. This structure of an algebra over the "Plumbers' PROP" is currently under construction by Yash Deshmukh. It is important to point out that formulating descent for this properadic structure is considerably more challenging. Let us be content with a simple example. Consider a presheaf of coalgebras whose underlying presheaf of modules satisfy the sheaf property. Then, the coalgebra structure on a disjoint union of two subsets is not uniquely determined by the two pieces as coalgebras. This is in contrast with what happens for a sheaf of algebras.*

4. Descent with algebraic structures

Let $K_1, \ldots, K_M$ be compact subsets of M. We apply the Thom-Whitney construction from Proposition 2.9 to $\mathcal{O}$ being the KSV algebra and $\mathcal{A}^\cdot$ the semi-cosimplicial KSV-algebra

$$\mathcal{SC}^\bullet := p \mapsto \bigoplus_{|J|=p+1} SC^*_{M,big}(\bigcap_{m\in J} K_m),$$

to obtain another KSV algebra

$$TW(SC^*_{M,big}; K_1, \ldots, K_M).$$

Theorem 4.1. *Assume that $K = K_1 \cup \ldots \cup K_N$ is a weakly involutive cover. Then, the canonical map*

$$SC^*_{M,big}(K) \to TW(SC_{M,big}; K_1, \ldots, K_N) \tag{9}$$

is a quasi-isomorphism of KSV-algebras.

Proof. By construction, the map respects the KSV structures (recall Proposition 2.9), so all we need to show is that the map induces an isomorphism on homology. It follows from Theorem 3.7 that we have a commutative diagram

$$\begin{array}{ccc} SC^*_M(K) & \longrightarrow & \check{\mathrm{C}}\mathrm{ech}(SC_M; K_1, \ldots, K_N) \\ \downarrow & & \downarrow \\ SC^*_{M,big}(K) & \longrightarrow & \check{\mathrm{C}}\mathrm{ech}(SC_{M,big}; K_1, \ldots, K_N) \end{array}$$

with vertical arrows being quasi-isomorphisms. Theorem 3.5 says that the upper arrow is a quasi-isomorphism and therefore the lower arrow is a quasi-isomorphism.

Moreover, by Propositions 2.6 and 2.7, we have a canonical commutative diagram

$$\begin{array}{ccc} & SC^*_{M,big}(K) & \\ \swarrow & & \searrow \\ TW(SC_{M,big}; K_1, \ldots, K_N) & \longrightarrow & \check{\mathrm{C}}\mathrm{ech}(SC_{M,big}; K_1, \ldots, K_N) \end{array}$$

with the horizontal arrow a quasi-isomorphism. This finishes the proof. □

Corollary 4.2. *Assume that $K = K_1 \cup \ldots \cup K_N$ is an involutive cover. Then, the canonical map*

$$SC^*_{M,big}(K) \to TW(SC_{M,big}; K_1, \ldots, K_N) \tag{10}$$

is a BV_∞ quasi-isomorphism of homotopy BV-algebras.

5. Mirror symmetry

We now explain with lightning speed a very long program that is aimed at a conceptualization of mirror symmetry. Our goal is to highlight the role played by the local-to-global principles of the last section. We do not touch upon homological mirror symmetry. The whole section should be thought of as conjectural.

Let us now assume that M is closed and graded with a weakly involutive cover $M = \bigcup_{i=1}^N C_i$, e.g. we have in mind Remark 2.3 with the assumption that the pair of the total space and the special fiber form a log CY pair [37].

In the mirror side, which is in the world of rigid analytic geometry [38, 39] for our purposes here, we will consider a smooth (meaning that the tangent sheaf TY [40, Section 9.1] is locally free) rigid analytic space Y over Λ with an admissible affinoid cover $Y = \bigcup_{i=1}^N Y_i$ and a global non-vanishing section of the canonical bundle $\bigwedge^n T^*Y$. By definition, $TY(Y_J)$ is the Lie algebra $\mathrm{Der}(\mathcal{O}_J)$ of Λ-linear derivations $O_J \to O_J$ (automatically bounded), where $\mathcal{O}_J$ is the algebra of functions on the affinoid domain $Y_J := \bigcap_{m \in J} Y_m$ for non-empty $J \subset \{1, \ldots, N\}$. Note that we have a BV (and hence a BV_∞) structure on $Sym^*_{\mathcal{O}_J}(\mathrm{Der}(\mathcal{O}_J)[1])$, where $Sym^*_{\mathcal{O}_J}$ stands for graded symmetric power [41, Section 2.1].

We can now assume that mirror symmetry holds locally and indicate how to deduce a global form of mirror symmetry as an application of the local-to-global principles. We state this as a conjecture being as precise as we can.

Conjecture 1. *Assume that for all non-empty $J \subset \{1, \ldots, N\}$, we have a BV_∞ quasi-isomorphism of homotopy BV-algebras*

$$SC^*_{M,big}\Big(\bigcap_{m \in J} C_m; \Lambda\Big) \to Sym^*_{\mathcal{O}_J}(Der(\mathcal{O}_J)[1]) \tag{11}$$

compatibly with respect to restriction maps. Then, we have a BV_∞ quasi-isomorphism of homotopy BV-algebras

$$SC^*_{M,big}(M;\Lambda) \simeq C^*(M;\Lambda) \to TW(Y, \bigwedge TY),$$

where $\bigwedge TY$ denotes the sheaf of polyvector fields and we are applying the Thom-Whitney construction with respect to $Y = \bigcup_{i=1}^N Y_i$ in the right hand side.

Remark 5.1. *We are being vague about what it means for the local BV_∞ quasi-isomorphisms in Equation* (11) *to be compatible with restriction maps, as the relevant notion of BV_∞ homotopies has not been sufficiently developed in the literature. Nevertheless, a compatibility of this form is a non-trivial condition. It is best to consider an example following* [42, Section 1.2]. *For the Thurston manifold with an involutive cover lifted from a good cover of the base of its Lagrangian torus fibration without a Lagrangian section and a non-archimedean abelian variety (with an appropriate cover) as the mirror, one can find local BV_∞ quasi-isomorphisms as in Equation* (11) *but cannot make these compatible with restriction maps altogether. The problem can be solved by equipping the rigid analytic space with a gerbe but we will not discuss this further.*

We could have written a statement with a complete proof if instead of homotopy BV-algebras and BV_∞ quasi-isomorphisms we used KSV algebras (by formality, BV algebras can be functorially turned to KSV algebras) and maps of KSV algebras. This would be a much less useful statement, as one cannot hope to produce quasi-isomorphisms as in Equation (11) *which strictly respect the KSV actions.*

Given a chain complex A with a homotopy BV algebra structure and a null-homotopy of the circle action, we obtain a hypercommutative algebra structure on $H(A)$ [43, 44]. Incorporating the cyclic (in the operadic sense) structures in these results, we recover the full genus 0 cohomological field theories [41, 45]. This is the main reason why it is important to state these results at the chain level. The discussion of these two extra pieces of structure is beyond the scope of this note.

The direct homology level implication (see [38, Theorem 5, Section 6.2] for the fact that we can compute sheaf cohomology using our affinoid cover) of Theorem 1 is that we have an isomorphism of BV-algebras

$$QH^*(M;\Lambda) \to H^*(Y, \bigwedge TY).$$

This is weaker than it might appear at first sight since it is known that the BV operators vanish on both sides (assuming Y is also proper, which is typical here, see Remark 5.4). Nevertheless, it says that the mirror of the small quantum product is the exterior product structure on the sheaf cohomology of polyvector fields, which is worth recording as a seperate statement. We make unrealistic assumptions and give a full proof.

Theorem 5.2. *We pick a point on $f\overline{\mathcal{M}}^{\mathbb{R}}_{0,2+1}$ and consider $SC^*_{M,big}(K;\Lambda)$'s only with the corresponding product structure.*

Assume that for all non-empty $J \subset \{1, \ldots, N\}$, we have a quasi-isomorphism of chain complexes

$$SC^*_{M,big}(\bigcap_{m \in J} C_m; \Lambda) \to Sym^*_{\mathcal{O}_J}(Der(\mathcal{O}_J)[1]) \tag{12}$$

compatible with the algebra structures on the nose. Then, we have an isomorphism of algebras

$$QH^*(M; \Lambda) \to H^*(Y, \bigwedge TY),$$

where $\bigwedge TY$ denotes the sheaf of polyvector fields.

Proof. Using the formula from [22, Equation (14.24)], we equip the relevant Čech complexes with a product. We can identify $QH^*(M; \Lambda)$, which is isomorphic to $SC^*_{M,big}(M; \Lambda)$ as an algebra, with the cohomology of Čech$(SC_{M,big}; C_1, \ldots, C_N)$ as an algebra using Theorem 3.5, cf. page 175 of [22]. We can do this on the mirror side as well. By the various compatibilities we are given, the map induced on Čech complexes is a map of algebras, which finishes the proof. □

Remark 5.3. *One might naively think that if the maps of Equation* (12) *are compatible with the restriction maps and the product structure not on the nose but only at the homology level, the conclusion of Theorem 5.2 would stay true. This is not correct because the closed elements of the Čech complexes are not built solely from closed elements of its summands. Even if we assume that the compatibility with restriction maps is on the nose, homology level compatibility with the product does not seem sufficient. We are not aware of a notion weaker than compatibility with the product on the nose that would allow us to conclude. A more realistic version of this statement would also assume that the maps of Equation* (12) *are compatible with the restriction maps only up to coherent homotopies. In this case, to conclude, one would also need compatibility of these homotopies with the product structure.*

It is well-known that the algebra structure on $SC^*_{M,big}(K; \Lambda)$ can be upgraded to an A_∞-structure. We can do this in such a way that restriction maps strictly respect these structures. It seems plausible that assuming the maps of Equation (12) *to be A_∞-quasi-isomorphisms which respect the restriction maps up to coherent homotopies of A_∞-maps (which can be formulated using the viewpoint of* [46, Remark 1.11]*) would do the job. On the other hand, at least at first glance, the coherence data related to the lack of associativity seems irrelevant for what we are trying to do.*

Remark 5.4. *Typically the rigid analytic spaces that appear in these statements will be proper (for example their polyvector field cohomologies will be finite dimensional) because we restricted ourselves to closed symplectic manifolds. The story extends to certain open and geometrically bounded symplectic manifolds as well, see* [47]*. Mirrors to specially chosen finite volume domains of such symplectic manifolds would fall into the framework above without being proper.*

We have a geometric context in which we expect to be able to construct Y from $M = \bigcup_{i \in I} C_i$ by gluing the affinoid domains

$$Y_i := MaxSpec(HF^*_M(C_i; S; \Lambda)),$$

for some reference Lagrangian submanifold S. The reference Lagrangian S is an abstraction of a Lagrangian section of an SYZ fibration, and it satisfies conditions that lead to the algebras $HF^*_M(C_i; S; \Lambda)$ being affinoid (in particular commutative) algebras supported in degree 0 as well as a certain local generation criterion. The latter means that

$$HH_{-n}(CF^*_M(C_i; S; \Lambda)) \to SH^0_M(C_i; \Lambda) \tag{13}$$

hits the unit for all $1 \leq i \leq N$.

The smoothness of Y and the local statements highlighted in Equation (11) are then expected consequences of the local generation package involving Cardy relations and more [48, 49, 50]; along with souped-up versions of the classical HKR theorem [51, 52, 53, 54, 55, 56].

Acknowledgements: The author thanks Bertrand Toën for a useful conversation about Section 2.3, Mohammed Abouzaid for explaining the role of gerbes in Section 5 and the referee(s) for useful feedback. U.V. was partially supported by the TÜBİTAK 2236 (CoCirc2) programme with a grant numbered 121C034 and by the Science Academy's Young Scientist Awards Program (BAGEP) in Turkey.

References

[1] Y. Groman, *Floer theory and reduced cohomology on open manifolds.* Geometry & Topology, **27(4)** (2023), 1273-1390.

[2] U. Varolgunes, *Mayer–Vietoris property for relative symplectic cohomology.* Geometry & Topology, **25** (2021), 547–642.

[3] A. Floer and H. Hofer, *Symplectic homology I: open sets in* C^n. Math. Z., **215** (1994), 37–88.

[4] K. Cieliebak, A. Floer, and H. Hofer, *Symplectic homology II: A general construction.* Math. Z., **218** (1995), 103–122.

[5] S. Venkatesh, *The quantitative nature of reduced Floer theory.* Adv. Math., **383** (2021), 107682.

[6] M. McLean, *Birational Calabi–Yau manifolds have the same small quantum products.* Annals of Mathematics, **191(2)** (2020), 439–579.

[7] K. Cieliebak and A. Oancea, *Symplectic homology and the Eilenberg–Steenrod axioms.* Algebr. Geom. Topol., **18(4)** (2018), 1953–2130.

[8] M. Abouzaid, Y. Groman, and U. Varolgunes, *Framed* E_2 *structures in Floer theory.* Adv. Math., **450** (2024), 109755.

[9] S. Piunikhin, D. Salamon, and M. Schwarz, *Symplectic Floer–Donaldson theory and quantum cohomology.* Contact and Symplectic Geometry (Cambridge, 1994), **8** (1996), 171–200.

[10] U. Varolgunes, *Mayer–Vietoris property for relative symplectic cohomology.* PhD thesis, MIT, (2018).

[11] C. Y. Mak, Y. Sun, and U. Varolgunes, *A characterization of heaviness in terms of relative symplectic cohomology.* Journal of Topology, **17(1)** (2024), e12327.

[12] R. Friedman, *Global smoothings of varieties with normal crossings.* Annals of Mathematics (2), **118** (1983), 75–114.

[13] M. F. Tehrani, M. McLean, and A. Zinger, *Singularities and semistable degenerations for symplectic topology.* C. R. Math. Acad. Sci. Paris, **356** (2018), 420–432.

[14] M. Harada and K. Kaveh, *Integrable systems, toric degenerations and Okounkov bodies.* Invent. Math., **202** (2015), 927–985.
[15] J. Fernández de Bobadilla and T. Pelka, *Fibrations by lagrangian tori for maximal Calabi–Yau degenerations and beyond.* arXiv:2312.13248, 2023.
[16] I. Kříž and J. P. May, *Operads, algebras, modules and motives.* Astérisque, no. 233, (1995).
[17] T. Kimura, J. Stasheff, and A. Voronov, *On operad structures of moduli spaces and string theory.* Comm. Math. Phys., **171** (1995), 1–25.
[18] E. Getzler, *Batalin–Vilkovisky algebras and two-dimensional topological field theories.* Comm. Math. Phys., **159** (1994), 265–285.
[19] J. Giansiracusa and P. Salvatore, *Formality of the framed little 2-discs operad and semidirect products.* Homotopy theory of function spaces and related topics, **519** (2010), 115–121.
[20] P. Ševera, *Formality of the chain operad of framed little disks.* Lett. Math. Phys., **93** (2010), 29–35.
[21] I. Gálvez-Carrillo, A. Tonks, and B. Vallette, *Homotopy Batalin–Vilkovisky algebras.* J. Noncommut. Geom., **6** (2012), 539–602.
[22] R. Bott and L. W. Tu, *Differential forms in algebraic topology.* Graduate Texts in Mathematics, vol. 82, Springer, (1982).
[23] D. Sullivan, *Infinitesimal computations in topology.* Publ. Math. IHÉS, **47** (1977), 269–331.
[24] V. Navarro Aznar, *Sur la théorie de Hodge–Deligne.* Invent. Math., **90** (1987), 11–76.
[25] X. Z. Cheng and E. Getzler, *Transferring homotopy commutative algebraic structures.* J. Pure Appl. Algebra, **212** (2008), 2535–2542.
[26] M. Hovey, *Model categories.* Math. Surveys Monographs, Amer. Math. Soc., **63** (1999).
[27] S. Arkhipov and S. Ørsted, *Homotopy (co)limits via homotopy (co)ends in general combinatorial model categories.* Appl. Categ. Struct., **31** (2023), 47.
[28] P. Griffiths and J. Morgan, *Rational homotopy theory and differential forms.* Prog. Math., vol. 16, Birkhäuser, (1981).
[29] V. Hinich, *Homological algebra of homotopy algebras.* Comm. Algebra, **25** (1997), 3291–3323.
[30] D. Tonkonog and U. Varolgunes, *Super-rigidity of certain skeleta using relative symplectic cohomology.* J. Topol. Anal., **15** (2023), 57–105.
[31] D. Salamon, *Lectures on Floer homology.* Symplectic Geometry and Topology, IAS/Park City Math. Ser., vol. 7, (1999), 143–229 .
[32] J. Hollender and R. Vogt, *Modules of topological spaces, applications to homotopy limits and E_∞-structures.* Arch. Math., **59** (1992), 115–129.
[33] K. Arakawa, *Homotopy limits and homotopy colimits of chain complexes.* arXiv:2310.00201, 2023.
[34] P. Seidel, *Fukaya categories and deformations.* Proc. ICM 2002 (Beijing), Higher Educ. Press, 351–360.
[35] T. Perutz and N. Sheridan, *Constructing the relative Fukaya category.* arXiv:2203.15482, 2022.
[36] C. Roger, *Gerstenhaber and Batalin–Vilkovisky algebras; algebraic, geometric, and physical aspects.* Archivum Math., **45** (2009), 301–324.
[37] M. Gross and B. Siebert, *The canonical wall structure and intrinsic mirror symmetry.* Invent. Math., **229** (2022), 1101–1202.
[38] S. Bosch, *Lectures on formal and rigid geometry.* LN Math., vol. 2105, Springer, (2014).
[39] J. Fresnel and M. van der Put, *Rigid analytic geometry and its applications.* Springer, (2012).
[40] K. Ardakov and S. J. Wadsley, *D–modules on rigid analytic spaces I.* J. Reine Angew. Math., **747** (2019), 221–275.
[41] S. Barannikov and M. Kontsevich, *Frobenius manifolds and formality of Lie algebras of polyvector fields.* Int. Math. Res. Not., **(4)**(1998), 201–215.
[42] M. Abouzaid, *Family Floer cohomology and mirror symmetry.* arXiv:1404.2659, 2014.
[43] G. C. Drummond–Cole and B. Vallette, *The minimal model for the Batalin–Vilkovisky operad.* Selecta Math., **19** (2013), 1–47.

[44] A. Khoroshkin, N. Markarian, and S. Shadrin, *Hypercommutative operad as a homotopy quotient of* BV. Comm. Math. Phys., **322** (2013), 697–729.
[45] Y. I. Manin, *Frobenius manifolds, quantum cohomology, and moduli spaces.* Amer. Math. Soc., (1999).
[46] P. Seidel, *Fukaya categories and Picard–Lefschetz theory.* EMS, (2008).
[47] Y. Groman and U. Varolgunes, *Closed string mirrors of symplectic cluster manifolds.* arXiv:2211.07523, 2022.
[48] M. Abouzaid, *A geometric criterion for generating the Fukaya category.* Publ. Math. IHÉS, **112** (2010), 191–240.
[49] S. Ganatra, *Symplectic cohomology and duality for the wrapped Fukaya category.* arXiv:1304.7312, 2013.
[50] S. Ganatra, *Cyclic homology,* S^1*-equivariant Floer cohomology, and Calabi–Yau structures.* arXiv e-prints (2019).
[51] G. Hochschild, B. Kostant, and A. Rosenberg, *Differential forms on regular affine algebras.* Trans. Amer. Math. Soc., **102** (1962), 383–408.
[52] V. Dolgushev, D. Tamarkin, and B. Tsygan, *The homotopy Gerstenhaber algebra of Hochschild cochains of a regular algebra is formal.* J. Noncommut. Geom., **1** (2007), 1–25.
[53] D. Calaque and M. Van den Bergh, *Hochschild cohomology and Atiyah classes.* Adv. Math., **224** (2010), 1839–1889.
[54] T. Willwacher and D. Calaque, *Formality of cyclic cochains.* Adv. Math., **231** (2012), 624–650.
[55] D. Calaque, C. A. Rossi, and M. Van den Bergh, *Calderaru's conjecture and Tsygan's formality.* Annals of Mathematics, **175 (2)**, (2012), 865–923.
[56] V. Dolgushev, D. Tamarkin, and B. Tsygan, *Formality of the homotopy calculus algebra of Hochschild (co)chains.* arXiv:0807.5117, 2008.

Koç University Mathematics Department, Rumelifeneri/İstanbul, TÜRKİYE
Email address: uvarolgunes@ku.edu.tr

Proceedings of 29th Gökova
Geometry-Topology Conference
pp. 37 – 56

Construction of symplectic flexible links

Johan Björklund, Georgios Dimitroglou Rizell

Abstract. We show that any smooth one-dimensional link in the real projective three-plane is the fixed-point locus of a smooth symplectic surface in the complex projective three-plane which is invariant under complex conjugation. The degree of the surface can be taken to be either one or two, depending on the homology class of the link. In other words, there are no obstructions to finding a symplectic representative of a flexible link beyond the classical topology.

1. Introduction and results

A smooth link $L \subset \mathbb{R}P^3 \subset \mathbb{C}P^3$ is called **real algebraic** if it is the real locus of a closed real smooth algebraic curve $\Sigma \subset \mathbb{C}P^3$, i.e. an embedded algebraic surface that is invariant under complex conjugation $I\colon \mathbb{C}P^3 \to \mathbb{C}P^3$, i.e. the real part of an embedded algebraic curve that is invariant under complex conjugation. For any submanifold $\Sigma \subset \mathbb{C}P^n$ which is invariant under complex conjugation I, we write $\Sigma_{\mathbb{R}} = \Sigma \cap \mathbb{R}P^3 = L$ for its fixed-point locus. Recall that I is an anti-symplectic involution for the standard Fubini–Study symplectic form ω_{FS}, that is $I^*\omega_{FS} = -\omega_{FS}$, and that the fixed-point locus of I is precisely $\mathbb{R}P^3$. An I-invariant surface $\Sigma \subset \mathbb{C}P^3$ is said to be of **Type-I** if $\Sigma \setminus \Sigma_{\mathbb{R}}$ is disconnected, and is otherwise said to be of **Type-II**.

The first author classified rational real algebraic knots up to degree five in [1]. The cases of degree less than six, and genus at most one, were later classified by Mikhalkin–Orevkov in [2]. Also, see work by Akbulut [3] for the existence of complexifications of knots (that typically live in high degrees). Real algebraic knots are naturally very rigid objects. For instance, it is easy to show that the isotopy class of $\Sigma_{\mathbb{R}}$ is unique when Σ is of degree one or two, even up to isotopy through algebraic curves. On the contrary, dropping the condition on Σ being algebraic and only requiring the surface to be orientable, I-invariant, and compatible with the complex orientation, one obtains a much more flexible type of object that was studied by the first author in [4]. Such a link in $\mathbb{R}P^3$ is called a **flexible link** and, as the name suggests, flexible links are determined by classical topological invariants and the knot type up to flexible isotopy; see the aforementioned work.

Key words and phrases. Real algebraic knots and links, flexible knots and links, anti-symplectic involution, symplectic submanifolds, transverse knots.

The author is supported by the Knut and Alice Wallenberg Foundation through the grants KAW 2021.0191 and KAW 2023.0294.

Compared to the real algebraic case, there is a big difference in the flexible case; namely, in this case the real locus does not determine the I-invariant surface, not even its topology. For this reason we also include the choice of the I-invariant surface itself as a part of the definition of a flexible link. In this note we are concerned with the following more restrictive subclass:

Definition 1.1. A **flexible-symplectic link** is an embedded symplectic I-invariant surface $\Sigma \subset (\mathbb{C}P^3, \omega_{FS})$ of real dimension $\dim \Sigma = 2$. We denote its fixed-point set by $\mathbb{R}\Sigma = \Sigma \cap \mathbb{R}P^3$, which is a smooth link.

The reason why we again call this object "flexible" is that the existence of an extension of a link $L \subset \mathbb{R}P^3$ to an I-invariant symplectic surface only is governed by classical topology. Establishing this fact is the main goal of our paper. Before stating the result, we point out the obvious fact that any closed symplectic surface $\Sigma \subset \mathbb{C}P^3$ must have a positive degree $d > 0$, i.e. must be homologous to $[\Sigma] = d[\mathbb{C}P^1] \in \mathbb{Z}[\mathbb{C}P^1] = H_2(\mathbb{C}P^3)$ with $d > 0$.

Theorem 1.2. *Any smooth link $L \subset \mathbb{R}P^3$ can be realised as the real-part $L = \Sigma_{\mathbb{R}}$ of a connected flexible-symplectic link of degree $d \in \{1, 2\}$ of the same parity as $[L] \in H_1(\mathbb{R}P^3) = \mathbb{Z}_2$. Furthermore, the surface $\Sigma \setminus L$ can be either of Type I or II; in the first case the genus can assume any value $g = \operatorname{rk} H_0(L) - 1 + 2k$ with $k \geq 0$, while in the second case it can assume any value $g = \operatorname{rk} H_0(L) - 1 + k$ with $k \geq 1$.*

We plan to study the classification question in future work, i.e. the set of flexible-symplectic links up to the relation of isotopy through links of the same type.

Remark 1.3. Any I-invariant surface $\Sigma \subset \mathbb{C}P^3$ of real dimension $\dim \Sigma = 2$ that intersects $\mathbb{R}P^3$ cleanly must have a degree of the same parity as that of the corresponding class in $[\Sigma \cap \mathbb{R}P^3] \in H_1(\mathbb{R}P^3) = \mathbb{Z}_2$. Namely, the divisor $\mathbb{C}P^2_\infty \subset \mathbb{C}P^3$ is fixed set-wise by I with fixed-point locus $\mathbb{R}P^2_\infty \subset \mathbb{R}P^3$, and the degrees of a curve in $\mathbb{C}P^3$ and $\mathbb{R}P^3$ are given by the intersection number with $\mathbb{C}P^2_\infty$ and $\mathbb{R}P^2_\infty$, respectively.

2. Background on symplectic and contact geometry

2.1. Symplectic manifolds

A **symplectic manifold** is a pair (X^{2n}, ω) consisting of a smooth even-dimensional manifold X^{2n} together with a closed non-degenerate two-form $\omega \in \Omega^2(X)$, i.e. a **symplectic two-form**.

Example 2.1. Here we are mainly interested in the complex projective space $(\mathbb{C}P^n, \omega_{FS})$ equipped with the Fubini–Study Kähler form $\omega_{FS} \in \Omega^{(1,1)}(\mathbb{C}P^n)$. This symplectic manifold has an integrable complex structure $J \in End(T\mathbb{C}P^n)$, $J^2 = -\mathrm{Id}$, and the the symplectic form is a Kähler form which in local affine coordinates can be written as

$$\omega_{FS} = \frac{i}{2}\partial\bar{\partial} \log\left(1 + |\mathbf{z}|^2\right) = -\frac{1}{4} dd^c \log\left(1 + |\mathbf{z}|^2\right), \quad d^c f := df \circ J,$$

i.e. it has a locally defined Kähler potential $\log\left(1+|\mathbf{z}|^2\right)$ in each standard affine coordinate patch. Note that the area of a line $\mathbb{C}P^1 \subset \mathbb{C}P^n$ is equal to $\int_{\mathbb{C}P^1} \omega_{FS} = \pi$ with our conventions.

A **symplectomorphism** is a diffeomorphism between symplectic manifolds that preserves the symplectic forms. A smooth isotopy ϕ^t through symplectomorphisms is a **Hamiltonian isotopy** if $\iota_{\dot{\phi}^t}\omega = -dH_t$ for some smooth function $H_t\colon X \to \mathbb{R}$ which is uniquely defined up to the addition of a locally constant function. The time-one map of a Hamiltonian isotopy is called a **Hamiltonian diffeomorphism**.

Example 2.2. The group of Kähler isometries of $(\mathbb{C}P^n, \omega_{FS})$ is equal to the projective unitary group $PU(n+1)$, which acts linearly on the homogeneous coordinates $\mathbb{C}^{n+1} \setminus \{0\}$ by elements in $U(n+1)$. Since these groups are connected and $\mathbb{C}P^n$ is simply connected, the elements consist of Hamiltonian diffeomorphisms.

A submanifold $\iota\colon Y \hookrightarrow (X, \omega)$ is said to be **symplectic** if the two-form $\iota^*\omega$ is symplectic. A standard fact about isotopies of symplectic submanifolds that will be crucial for us is the "Hamiltonian Thom Isotopy Lemma", which is a folklore result that was given a thorough proof by Antonini–Cavalletti–Lerario in [5]:

Theorem 2.3 (The Hamiltonian Thom Isotopy Lemma; e.g. [5])**.** *A smooth isotopy $\iota_t\colon (Y, \iota_t^*\omega) \hookrightarrow (X, \omega)$ of symplectic embeddings which is fixed outside of a compact subset of the domain is generated by a compactly supported Hamiltonian isotopy, in the sense that $\iota_t(Y) = \phi^t \circ \iota_0(Y)$, where ϕ^t is a compactly supported Hamiltonian isotopy of (X, ω).*

Remark 2.4.

1. The proof in [5] is for isotopies of compact symplectic submanifolds of compact symplectic manifolds. However, the proof immediately works also for compactly supported isotopies of properly embedded symplectic manifolds. The latter version of the result translates to the version that we stated above.
2. Since we only make requirements about the images $\phi^t \circ \iota_0(Y) = \iota_t(Y)$, and not the parametrised maps, there is no need to assume that Y is simply connected. Note that the two parametrised symplectic embeddings $\phi^t \circ \iota_0$ and ι_t of the domain (Y, ω_Y) thus very well may differ by a non-Hamiltonian symplectic isotopy of Y.

Recall that a submanifold $L \subset (X^{2n}, \omega)$ is said to be **isotropic** if ω pulls back to zero on L. This implies that L is at most half-dimensional. If the dimension, moreover, is equal to half i.e. $\dim L = n$, then we say that L is **Lagrangian**.

Example 2.5. The one-form $d^c \log\left(1+|\mathbf{z}|^2\right)$ vanishes along the real part $\mathbb{R}P^n \subset \mathbb{C}P^n$ in each standard affine coordinate patch. In other words, $\mathbb{R}P^n \subset (\mathbb{C}P^n, \omega_{FS})$ is a Lagrangian submanifold.

When I is an **anti-symplectic involution** of (X, ω), i.e. $I^*\omega = -\omega$, it follows that the fixed-point set of I is a Lagrangian submanifold.

Example 2.6. The pull-back of the Fubini–Study form under complex conjugation is equal to

$$I^* \frac{i}{2} \partial\overline{\partial} \log\left(1 + |\mathbf{z}|^2\right) = \frac{i}{2} \overline{\partial}\partial \log\left(1 + |\mathbf{z}|^2\right) = -\omega_{FS},$$

since the operators ∂ and $\overline{\partial}$ anti-commute. In other words, I is an anti-symplectic involution. Since $\mathbb{R}P^n$ is the set of fixed-points of this involution, this also implies that $\mathbb{R}P^n$ is Lagrangian.

2.2. Contact manifolds

Contact geometry can be expressed in terms of conical symplectic geometry. More precisely, a **contact manifold** is a pair (Y^{2n+1}, ξ) that consists of an odd-dimensional manifold together with a maximally non-degenerate field of tangent hyperplanes $\xi \subset TY$. We will be interested in the case when $\xi = \ker \alpha$ for some auxiliary contact one-form $\alpha \in \Omega^1(Y)$. The contact condition can be expressed as the condition that the conical manifold

$$(\mathbb{R}_\tau \times Y, d(e^\tau \alpha))$$

is symplectic.

A one-dimensional knot or link in a contact manifold $T \subset (Y, \xi)$ is said to be **transverse** if it is everywhere transverse to the contact hyperplane distribution $\xi \subset TM$. This condition can be equivalently formulated as the condition that

$$\mathbb{R} \times T \subset (\mathbb{R}_\tau \times Y, d(e^\tau \alpha))$$

is a symplectic cone. An oriented knot is said to be **positively transverse** if α is positive on its oriented tangent line. The high-dimensional analogue of a transverse submanifold is a submanifold $T \subset (Y, \alpha)$ onto which α restricts to a contact form. In particular, the h-principle for iso-contact embeddings of high codimension implies that

Proposition 2.7 ([6]). *Two positively transverse one-dimensional links in a $(2n-1)$-dimensional contact manifold with $n \geq 3$, say $L_i \subset (Y^{2n-1}, \alpha)$ for $i = 0, 1$, are isotopic through transverse links if and only if they are homotopic.*

Furthermore, when $L_i := T_i \sqcup I_Y(T_i)$, $i = 0, 1$, for some fixed-point free contact involution

$$I_Y \colon (Y^{2n-1}, \alpha) \to (Y^{2n-1}, -\alpha),$$

and if T_0 and T_1 are homotopic, then there exists a transverse isotopy from L_0 to L_1 through I_Y-invariant links.

Proof. The first part is a direct application of the h-principle for iso-contact immersions [6, 24.1.2].

To construct the I_Y-equivariant transverse isotopy we argue as follows. First use the aforementioned h-principle to create a transverse isotopy T_t from T_0 to T_1. It is then enough to perturb this isotopy to one which is in general position, to ensure that $T_t \sqcup I_Y(T_t)$ remains embedded for all t. □

An $(n-1)$-dimensional submanifold $\Lambda \subset (Y,\xi)$ of a $(2n-1)$-dimensional contact manifold is **Legendrian** if it is tangent to the contact distribution ξ (this is the maximal possible dimension since ξ is maximally non-integrable by definition of the contact condition). Equivalently, this can be formulated as the condition that

$$\mathbb{R} \times \Lambda \subset (\mathbb{R}_\tau \times Y, d(e^\tau \alpha))$$

is a Lagrangian cone.

2.3. Cotangent bundles

Any cotangent bundle T^*M has the **tautological symplectic structure** defined by the exact two-form $d\theta_M$. Here $\theta_M \in \Omega^1(T^*M)$ is the **tautological one-form**, which has the following local expression. Given local coordinates (q^i, p_i) on T^*M, where q^i are any local coordinates on M and the p_i are the coordinates induced by q^i as the coefficients of the locally defined one-forms $p_i\, dq^i \in T^*M$, in other words p_i are locally defined conjugate momentum coordinates, we can write $\theta_M = \sum_i p_i\, dq_i$. The cotangent fibres T^*_pM as well as the zero-section $0_M \subset T^*M$ are Lagrangian with respect to the tautological symplectic structure. Furthermore, any diffeomorphism ϕ of M lifts to a symplectic bundle-morphism of $(T^*M, d\theta_M)$ through the pull-back $(\phi^{-1})^*$ of the inverse. In fact, $(\phi^{-1})^*\theta_M = \theta_M$, i.e. the diffeomorphisms act on T^*M while preserving the tautological one-form.

Any cotangent bundle $(T^*M, d\theta_M)$ has a canonical anti-symplectic involution I_0 given by fibre-wise multiplication by -1.

Given a compact subset $U \subset T^*M$ that is fibre-wise star-shaped with respect to the origin in the fibres, e.g. the radius-r co-disc bundle $D^*_{g,r}M$ for some Riemannian metric g on M, its boundary $Y = \partial U \subset T^*M$ is a contact manifold with contact form $\alpha := \theta_M|_{TY}$. Furthermore, the symplectisation of (Y,α) has a natural embedding in T^*M via

$$\begin{aligned}(\mathbb{R}_\tau \times Y, e^\tau \alpha) &\xrightarrow{\cong} (T^*M \setminus 0_M, \theta_M)\\ (\tau, y) &\mapsto e^\tau \cdot y,\end{aligned}$$

which preserves the primitives of the symplectic forms.

In the case when the star-shape subset U moreover is invariant under the anti-symplectic involution I, it follows that I_0 is induced by a co-orientation reversing contactomorphism without fixed points

$$I_Y \colon (Y, \ker\alpha) \to (Y, \ker\alpha)$$

that satisfies $I_Y^*\alpha = -\alpha$. Namely, I is given by extending I_Y to $T^*M \setminus 0_M \cong \mathbb{R}_\tau \times Y$ by $\mathrm{Id}_\mathbb{R}$ on the $\mathbb{R}_\tau$-factor.

2.4. Isotropic, Lagrangian, and symplectic cobordisms

It will be necessary to consider also cobordisms embedded in

$$(\mathbb{R}_\tau \times Y, d(e^t\alpha))$$

beyond simply trivial cones over submanifolds in Y. We say that a properly embedded submanifold C of the symplectisation has **conical ends** if it coincides with

$$(-\infty,-1] \times K_- \ \sqcup \ [1,+\infty) \times K_+ \subset (\mathbb{R}_\tau \times Y, d(e^t\alpha))$$

outside of a compact subset. In this case we say that C is a **cobordism from $K_- \subset Y$ to $K_+ \subset Y$ with conical ends**.

A cobordism C_0 from K_- to K, and a cobordism C_1 from K to K_+ can be concatenated in a manner to form a cobordism from K_- to K_+. This is done by first translating C_0 (resp. C_1) sufficiently far in the negative (resp. positive) $\mathbb{R}_\tau$-direction, so that its positive (resp. negative) end contains the level $\{\tau = 0\}$. After this has been done, we can form the concatenation in the following way:

Definition 2.8. After a translation as above, the union

$$C_0 \odot C_1 := C_0 \cap \{\tau \leq 0\} \ \cup \ C_1 \cap \{\tau \geq 0\}$$

is a cobordism from K_- to K_+ which is called the **concatenation** of C_0 and C_1.

We recall the classical Weinstein's Lagrangian neighbourhood theorem as well as a mild generalisation to the equivariant setting in the presence of an anti-symplectic involution.

Theorem 2.9 (Weinstein's Lagrangian Neighbourhood Theorem [7]). *Any proper Lagrangian embedding $\phi\colon L \hookrightarrow (X,\omega)$ can be extended to a symplectic embedding*

$$\Phi\colon (D^*L, d(p\,dq)) \hookrightarrow (X,\omega), \ \ \Phi|_{0_L} = \phi,$$

where D^*L is the unit cotangent bundle for some choice of Riemannian metric on L.

In addition,

1. *Let $(X,\omega) = (\mathbb{R}_\tau \times Y, d(e^\tau\alpha))$, and assume that there exists an embedding $\phi_\Lambda\colon \Lambda \hookrightarrow Y$ and coordinates $(s,x) \in I \times \Lambda \hookrightarrow L$, with $I \subset \mathbb{R}_s$ connected, such that $(\tau,y) = \phi(s,x) = (s,\phi_\Lambda(x))$. Using $\phi^t_{\partial_s}$ and $\phi^t_{\partial_\tau}$ to denote the time-t flow of the respective coordinate vector fields (i.e. translations of the $\mathbb{R}$-factors in the domain and target) we can assume that the symplectomorphism Φ above satisfies*
$$\Phi \circ (e^t \cdot (\phi^{-t}_{\partial_s})^*) = \phi^t_{\partial_\tau} \circ \Phi, \ \ on\ D^*(I\times\Lambda) \cap \{s+t \in I\},$$
i.e. translation in the symplectisation corresponds to translation and fibre-wise rescaling in the domain $D^*(I \times \Lambda) \subset D^*L$;
2. *If $I\colon (X,\omega) \to (X,-\omega)$ is an anti-symplectic involution that fixes L set-wise, but which is fixed-point free on L, then Φ can be assumed to satisfy the property that $\Phi^{-1} \circ I \circ \Phi$ coincides with the anti-symplectic involution $I_0 \circ (I|_L)^*\colon D^*L \to D^*L$.*
3. *When the conditions in (1) and (2) hold simultaneously, and moreover $I(\tau,y) = (\tau, I_Y(y))$ is satisfied (i.e. the anti-symplectic involution is induced by a contactomorphism of $(Y, \ker\alpha)$ that satisfies $I_Y^*\alpha = -\alpha$), then we may assume that the conclusion of (2) holds everywhere, simultaneously as the conclusion of (1) in the subset where the the cobordism is conical.*

Proof. The case without the additional assumptions is due to Weinstein [7].

(1): Since a conical Lagrangian cobordism is a cone over a Legendrian, this can be seen to follow from the strict standard Legendrian neighbourhood theorem [8, Theorem 6.2.2] by which any Legendrian $\Lambda \subset (Y, \alpha)$ has a neighbourhood which admits a contact-form preserving identification with a neighbourhood of the zero-section in $(J^1\Lambda = T^*\Lambda \times \mathbb{R}_z, dz + \theta_\Lambda)$. In the conical subset, one can use the fact that the cone over such a standard neighbourhood in the symplectisation is symplectomorphic to a neighbourhood of the zero-section $\mathbb{R} \times \Lambda \subset T^*(\mathbb{R} \times \Lambda)$.

(2): Since I is assumed to be without fixed-points near L, the second part can be readily proven by the same method as the standard version of Weinstein's Lagrangian neighbourhood theorem, but after making all choices I-invariant. The details are left to the reader.

(3): This follows from a combination of the techniques in the proofs of (1) and (2). □

Here we are mainly interested in cobordisms that are symplectic surfaces, in which case the ends are cones over transverse knots. However, to produce these symplectic cobordisms, we need to go pass via constructions of Lagrangian cobordisms, in which case the ends are cones over Legendrian submanifolds.

An isotopy $T_t \subset (Y, \xi)$ of transverse knots yields a symplectic version of the trace cobordism in the following way. We may assume that the isotopy T_t is constantly equal to $T_\pm$ whenever $\pm t \geq 1$.

Lemma 2.10. *The* ***symplectic trace cobordism*** *from T_- to T_+ given by*

$$\mathcal{T}_{\{T_t\}} = \{(\tau, y);\ y \in T_{v\tau}\} \subset (\mathbb{R} \times Y, d(e^\tau \alpha))$$

is symplectic if the speed $v > 0$ is taken to be sufficiently small. Furthermore,

- *When $Y = U^*M$ and T_t are I_Y-invariant, the trace cobordism is I_0-invariant as well; and*
- *The trace cobordism depends smoothly on the isotopy T_t.*

2.5. Symplectic push-offs of Lagrangian cobordisms

A two-dimensional Lagrangian cobordism can be pushed-off by a C^∞-small deformation to a symplectic cobordism between transverse knots. This was shown e.g. in [9, Lemma 4.1], and we recall the proof here. For completeness we give a completely general version of the construction that can be applied in any dimension.

Proposition 2.11. *Let $\phi\colon L \hookrightarrow (\mathbb{R}_\tau \times Y, d(e^\tau \alpha))$ be an even-dimensional Lagrangian cobordism from the Legendrian Λ_- to Λ_+. Assume that $\Lambda_\pm$ admit contact forms $\alpha_\pm \in \Omega^1(\Lambda_\pm)$ for which the induced one-form on L that coincides with $e^{\tau \circ \phi} \cdot \alpha_\pm$ on the positive and negative cylindrical ends, respectively, extends to a globally defined primitive $\eta \in \Omega^1(L)$ of a symplectic form $d\eta$. Then, there exists a smooth family of cobordisms*

$$\phi_t\colon L \hookrightarrow \mathbb{R} \times Y,\ \ \phi_0 = \phi,\ \ \phi_t^* e^\tau \alpha = t\eta,\ \ t \in [0, \epsilon),$$

with conical ends, for some small $\epsilon > 0$; hence $\iota_t(L)$ is symplectic for all $t \in (0, \epsilon)$.

Moreover, if $I(\tau, y) = (\tau, I_Y(y))$ is an $\mathbb{R}_\tau$- equivariant anti-symplectic involution of $(\mathbb{R} \times Y, d(e^\tau \alpha))$ that fixes L set-wise, has no fixed points on L, and satisfies $(I|_L)^*\eta = -\eta$, then ι_t can all be assumed to be I-invariant.

Remark 2.12. The one-form $\eta \in \Omega^1(L)$ makes (L, η) into a complete Liouville cobordism. In particular, L must be even-dimensional and have a non-empty convex conical end $\Lambda_+ \neq \emptyset$ holds by Stokes' theorem.

Proof. We choose a standard Weinstein neighbourhood

$$\Phi\colon D^*L \hookrightarrow (\mathbb{R} \times Y, d(e^\tau \alpha))$$

extending the Lagrangian embedding $\phi\colon L \hookrightarrow \mathbb{R} \times Y$ as in Theorem 2.9 that is compatible with the symplectic involution. The cobordism is then simply given by the image of the section of $t\eta$ in the cotangent bundle D^*L.

Moreover, when L is invariant under the anti-symplectic involution I, the sections $t\eta$ are invariant under the corresponding anti-symplectic involution $I_0 \circ (I|_L)^*$ on the cotangent bundle. □

3. Conical submanifolds of cotangent bundles

For constructing and deforming I-invariant submanifolds, it will be useful to first introduce the notion of a conical submanifold of a cotangent bundle.

Definition 3.1. Consider an immersion $\gamma : B \looparrowright M$ and a sub-bundle $E \to B$ of the pull-back bundle $\gamma^*(T^*M) \to B$. The natural immersion $E \looparrowright T^*M$ that lifts $B \looparrowright M$ is said to be a **conical** immersion, and we denote its image by $\mathcal{C}_E \subset T^*M$. If an immersion coincides with a conical immersion near 0_M, it is said to be **conical near** 0_M.

Note that conical submanifolds and immersions are invariant under I_0.

Example 3.2. For any submanifold $B \subset M = 0_M$ there are two natural conical extensions to T^*M with very different properties:

(1) For the conormal bundle of $B \subset M$, i.e. the bundle of cotangent vectors that annihilate TB, the corresponding conical submanifold in T^*M is Lagrangian. This conical submanifold is called the **conormal bundle of** B.
(2) There is a canonical bundle-isomorphism of $T^*M \to M$ and the vertical sub-bundle
$$V := \ker\left(T(T^*M)|_{0_M} \xrightarrow{T\pi} TM\right) \cong T^*M$$
as vector bundles over $M = 0_M \subset T^*M$. For any compatible almost complex structure J that interchanges the two sub-bundles $V, T0_M \subset T(T^*M)$ we get a symplectic conical submanifold given by the sub-bundle $J(TB) \subset T^*M$ along $B \subset 0_M$, i.e. an embedding of the cotangent bundle $T^*B \subset T^*M$ of B. This conical submanifold will be called the **cotangent extension of** B.

The construction also works for immersions $B \hookrightarrow M$, in which case it produces conical Lagrangian or symplectic immersions.

In order to better understand the conical submanifolds near $0_M \subset T^*M$, it is often useful to pass to the symplectisation. Clearly a conical immersion is $\mathbb{R}_\tau$-invariant inside the coordinates $\mathbb{R}_\tau \times Y \cong T^*M \setminus 0_M$, from which we conclude that:

Lemma 3.3. *Conical symplectic and Lagrangian immersions correspond to cones over submanifolds in the unit cotangent bundle that are transverse and Legendrian, respectively.*

3.1. Deformation of conical symplectic surfaces

For simplicity we here restrict our attention to two-dimensional conical symplectic submanifolds, which is the only case that we will need here. Naturally, many of these results have generalisations to conical submanifolds of higher dimension.

Lemma 3.4. *Let $\Sigma \subset T^*M$ be an I_0-invariant symplectic embedding of a surface of real dimension two that intersects 0_M cleanly in a one-dimensional link $L = \Sigma \cap 0_M$. Then Σ is Hamiltonian isotopic through surfaces that all are*

- *I_0-invariant, and*
- *intersect 0_M in L,*

to a surface that is conical near 0_M. The Hamiltonian isotopy can, moreover, be assumed to be supported in an arbitrarily small compact subset of 0_M.

Proof. Consider a conical embedding $\Sigma_0 \subset T^*M$ of $L \times \mathbb{R}$ that is tangent to Σ along L. Hence, Σ_0 is I_0-invariant and symplectic. Since C^1-small deformations of symplectic surfaces remain symplectic, it is simply a matter of perturbing Σ through I_0-invariant surfaces in a sufficiently small neighbourhood of L in order to make it coincide with Σ_0 there. We can then invoke Theorem 2.3 to produce the sought Hamiltonian isotopy. □

Lemma 3.5. *Let $\Sigma_i \subset T^*M$, $i = 0, 1$, be two symplectic embeddings of a surface of real dimension two, where both surfaces are conical near 0_M and intersects 0_M cleanly in the same one-dimensional link $L = \Sigma_i \cap 0_M$. Then Σ_0 can be made to agree with Σ_1 near 0_M by a Hamiltonian isotopy $\Sigma_t = \phi^t(\Sigma_0)$ of symplectic surfaces that has support in an arbitrarily small neighbourhood of 0_M. Furthermore, we can assume that Σ_t all symplectic surfaces that intersect 0_M in the same link L, and which all are conical near 0_M.*

Proof. Let $E_i \to L$, $i = 0, 1$, be the sub-bundle of $T^*M|_L \to L$ that corresponds to the conical symplectic submanifold that agrees with Σ_i near 0_M. The symplectic condition implies that E_i are complements of the conormal bundle. Since any two such complementary bundles can be deformed through sub-bundles that are complementary to the conormal bundle, we get a smooth family of conical symplectic surfaces $\mathcal{C}_{E_t,L}$ that connects $\mathcal{C}_{E_0,L}$ and $\mathcal{C}_{E_1,L}$.

The isotopy $\mathcal{C}_{E_t,L}$ of conical symplectic surfaces induces an isotopy of I_Y-invariant transverse knots $T_t \subset U^*M$; see Lemma 3.3. Using Lemma 2.10, we produce a one-parameter family of conical I_0-invariant symplectic trace-cobordisms $\mathcal{T}_t \subset \mathbb{R} \times U^*M$ from T_0 to T_t, with a smooth dependence on $t \in [0, 1]$. Furthermore $\mathcal{T}_0 = \mathbb{R} \times T_0$.

For the induced family of concatenations of $\mathcal{T}_t$ and $\mathcal{C}_{E_t,B}$ (see Definition 2.8), we obtain a smooth isotopy of symplectic surfaces that coincides with $\mathcal{C}_{E_t,B}$ near 0_M. The isotopy can moreover be assumed to be supported in an arbitrarily small neighbourhood of 0_M. Finally, the sought Hamiltonian isotopy is produced by invoking Theorem 2.3. □

3.2. Construction of two-dimensional symplectic cobordisms

The surfaces provided by Theorem 1.2 will be constructed by concatenations of standard pieces; see Definition 2.8. In this subsection we construct these needed standard pieces.

More precisely, we are interested in constructing two-dimensional symplectic cobordisms. This will be done by relying on constructions on exact two-dimensional Lagrangian cobordisms, which can be turned into symplectic surfaces through the push-off construction in Proposition 2.11.

However, two-dimensional Lagrangian cobordisms live inside an ambient four-dimensional symplectic manifold. Since we want to construct two-dimensional symplectic cobordisms inside six-dimensional symplectic manifolds, the surfaces that we construct inside a four-dimensional symplectic manifold will be embedded into six-dimensions via the stabilisation

$$(\mathbb{R}_\tau \times Y^3 \times \{0\}, d(e^\tau \alpha)) \hookrightarrow (\mathbb{R}_\tau \times Y^3 \times B^{2n}, d(e^\tau(\alpha + x\,dy))$$

which is the symplectisation of the contact product $(Y, \alpha) \times (B^{2n}, x\,dy)$, where $x\,dy = \sum_{i=1}^n x_i\,dy_i$ is a primitive of the standard linear symplectic form.

When we also are considering an anti-symplectic involution, we will assume that it is induced by a contact-involution

$$I_Y \times I_{\mathbb{C}^n} \colon (Y \times B^{2n}, \alpha + y\,dx) \to (Y \times B^{2n}, -\alpha - y\,dx),$$

which acts while preserving the two factors, and which is given by complex conjugation in the second factor.

Before we start with the constructions we will recall the following useful identification. Consider the standard contact form on the jet-space $J^1S^{k-1} = T^*S^{k-1} \times \mathbb{R}_z$ which is given by $\alpha_0 = dz - \theta_{S^{k-1}}$, where $\theta_{S^{k-1}} \in \Omega^1(T^*S^{k-1})$ is the tautological one-form. There is a fixed-point free contact involution

$$I_{J^1S^{k-1}} \colon (J^1S^{k-1}, dz - \theta_{S^{k-1}}) \to (J^1S^{k-1}, -(dz - \theta_{S^{k-1}})),$$
$$T^*S^{k-1} \times \mathbb{R} \ni (y, z) \mapsto \left(I_0^{T^*S^{k-1}} \circ A^*(y), -z\right),$$

where $A \colon S^{k-1} \to S^{k-1}$ is the antipodal map, and

$$I_0^{T^*S^{k-1}} \circ A^* \colon (T^*S^{k-1}, \theta_{S^{k-1}}) \to (T^*S^{k-1}, -\theta_{S^{k-1}})$$

is the anti-symplectic involution obtained by composing the symplectomorphism induced by the antipodal map with fibre-wise multiplication by -1.

Lemma 3.6 (Arnold's hodograph transformation [10]). *Consider the unit cotangent bundle with the contact form induced by the flat metric $(U^*\mathbb{R}^n, \alpha_0)$. For any linear subspace $V^k \subset T_0\mathbb{R}^n = \mathbb{R}^n$, the isotropic $k-1$-sphere $V \cap U_0^*\mathbb{R}^n$ has a neighbourhood that is strict* *contactomorphic (i.e. the contactomorphism preserves the contact forms and not merely the contact distributions) to a neighbourhood of* $j^1 0 \times \{0\}$ *inside*

$$(J^1 S^{k-1} \times T^*\mathbb{R}^{n-k}, dz - \theta_{S^{k-1}} + \theta_{\mathbb{R}^{n-k}})$$

under which

- *the neighbourhood intersected with* $U_0^*\mathbb{R}^n$ *is mapped into the zero-section* $j^1 0 \times 0_{\mathbb{R}^{n-k}}$*; and*
- *the contact-involution on* $U^*\mathbb{R}^n$ *corresponds to* $I_{J^1 S^{k}-1} \times I_{T^*\mathbb{R}^{n-k}}$*, where* $I_{T^*\mathbb{R}^{n-k}} = I_0 \circ (-\mathrm{Id}_{\mathbb{R}^{n-k}})^*$ *(i.e. multiplication by* -1 *in the base* $\mathbb{R}^{n-k}$*).*

Proof. Arnold's hodograph transformation takes care of the case $n = k$. The general case can be seen as follows.

The sphere $V^k \cap S^{n-1} \subset S^{n-1}$ has a neighbourhood $U \subset S^{n-1}$ that is diffeomorphic to the trivial bundle $S^{k-1} \times \mathbb{R}^{n-k} \to S^{k-1}$. Furthermore, under this identification, $V^k \cap S^{n-1}$ is identified with the zero-section, while the antipodal map of S^{n-1} becomes the antipodal map on the base S^{k-1} together with multiplication by -1 in the fibres $\mathbb{R}^{n-k}$ (i.e. in the normal bundle of the sphere).

Finally, note that there is a canonical identification

$$(J^1 U, dz - \theta_U) \cong (J^1 S^{k-1} \times T^*\mathbb{R}^{2(n-k)}, dz - \theta_{S^{k-1}} + \theta_{\mathbb{R}^{n-k}})$$

preserving the contact forms, and under which the involution takes the sought expression. □

Next, we give the recipes for constructing the needed two-dimensional I_0-invariant symplectic cobordisms inside the symplectisation $\mathbb{R} \times U^*M \cong T^*M \setminus 0_M$ whenever M is of dimension at least $\dim M \geq 2$.

3.2.1. Fillings of transverse knots by genus-g surfaces

Consider the standard Legendrian (p, q) torus link $\Lambda_{(p,q)} \subset (\mathbb{R}^3, dz - y\,dx)$ inside the standard contact vector space. There is a Lagrangian cobordism inside $\mathbb{R} \times \mathbb{R}^3$ from $\emptyset$ to $\Lambda_{p,q}$ which is of genus $\frac{1}{2}(p-1)(q-1)$; see e.g. [11]. Since every contact manifold is locally contactomorphic to the contact vector space, we can now apply Proposition 2.11 combined with the method of Subsection 3.2 to construct symplectic cobordisms of any genus from $\emptyset$ to the unique null-homotopic transverse knot in the contactisation of any contact manifold of dimension at least five. (Recall Proposition 2.7 by which there is a unique null-homotopic transverse knot in connected high-dimensional contact manifolds.)

Since the aforementioned cobordism can be constructed inside the symplectisation of any arbitrarily small neighbourhood of a contact manifold, it is easy to construct an I_0-invariant such cobordism consisting of two disjoint genus-g surfaces.

3.2.2. Connected I_0-invariant filling without fixed points

Here we construct a connected I_0-invariant filling in the symplectisation of J^1S^1 on which I_0 has no fixed points. First, consider the two-copy link

$$j^1(1) \sqcup j^1(-1) \subset (J^1S^1, dz - \theta_{S^1}) = (S^1 \times \mathbb{R}_p \times \mathbb{R}_z, dz - p\,d\theta),$$

which is invariant under the contact-involution $I_{J^1S^1}(\theta, p, z) = (\theta + \pi, -p, -z)$. This Legendrian has a connected Lagrangian $I_{J^1S^1}$-invariant filling in $\mathbb{R}_\tau \times J^1S^n$ given by

$$\{(\gamma_1(t), j^1\gamma_2(t)) \in \mathbb{R}_\tau \times J^1S^n\}$$

for some smooth curve $(\gamma_1, \gamma_2)(t) \in \mathbb{R}^2$ which satisfies

- $\gamma_1(t) = \gamma_1(-t)$;
- $\gamma_1''(t) \geq 0$; and
- $\gamma_1(t) = |t|$ for $t \gg 0$,

while

- $\gamma_2(t) = -\gamma_2(-t)$;
- $\gamma_2'(t) \geq 0$; and
- $\gamma_2(\pm t) = \pm 1$ for $t \gg 0$.

Note that $j^1\gamma_2(t) = \{z = \gamma_2(t)\} \subset J^1S^1$ is a push-off of the zero-section j^10 by the time-$\gamma_2(t)$ Reeb flow. Hence, the pull-back of $e^\tau(dz - p\,d\theta)$ under the map $(\gamma_1, j^1\gamma_2)$ is clearly an exact one-form, which means that this map is an exact Lagrangian embedding.

Again, since J^1S^1 embeds into the standard contact Darboux ball, we can argue as in Subsection 3.2.1 to produce a symplectic I_0-invariant cylinder in the symplectisation $\mathbb{R} \times U^*M$ whenever $\dim M \geq 2$.

3.2.3. Pairs of boundary connected sums via invariant pairs of one-handles

Here we describe how one can perform a Lagrangian boundary connected sum operation inside the symplectisation via a pair of Lagrangian one-handles that are exchanged under the involution I_0. Thus, both the one-handles and the connected sum operations are I_0-invariant. The construction is as follows:

One can perform the so-called Legendrian ambient connected-sum operation on two disjoint Legendrians $\Lambda_0 \sqcup \Lambda_1 \subset (Y, \alpha)$ as described in work [12] by the second author. Furthermore, there is a Lagrangian standard one-handle attachment cobordism from the union $\Lambda_0 \sqcup \Lambda_1$ to the result Λ_3 of the connected sum. Concatenating the cobordism $\mathbb{R} \times (\Lambda_0 \sqcup \Lambda_1)$ with this Lagrangian one-handle amounts to performing a boundary connected sum.

In the case when $\Lambda_0 \sqcup \Lambda_1 \cap I_Y(\Lambda_0 \sqcup \Lambda_1) = \emptyset$, we can use a sufficiently thin Lagrangian handle-attachment cobordism to ensure that also the cobordism is disjoint from its own image under the anti-symplectic involution I_0. This is the sought I_0-invariant pair of Lagrangian one-handles that give rise the two simultaneously performed boundary connected sums.

Finally, arguing as above, one can readily turn this construction into a boundary-connected sum operation on symplectic surfaces in symplectisations.

4. Conical surfaces in the complex projective space

For constructing I-invariant symplectic submanifolds of $\mathbb{C}P^n$ it is useful to symplectically identify a neighbourhood of the fixed-point locus $\mathbb{R}P^n \subset \mathbb{C}P^n$ of I with a neighbourhood of $0_{\mathbb{R}P^n} \subset T^*\mathbb{R}P^n$. There are many such identification, e.g. by Weinstein's neighbourhood theorem. We will follow a path similar to the one by Adaloglou [13], in order to obtain an identification of a neighbourhood as large as possible.

The standard projective plane $\mathbb{R}P^n \subset \mathbb{C}P^n$ is contained inside the complement

$$\mathbb{R}P^n \subset \mathbb{C}P^n \setminus Q^{n-1}$$

of the standard smooth quadric

$$Q^{n-1} = \{[Z_0 : Z_1 : \ldots : Z_n];\ Z_0^2 + Z_1^2 + \ldots + Z_n^2 = 0\}.$$

Note that the involution I fixes Q^{n-1} as a subset, and thus descends to a fixed-point free anti-symplectic involution on Q^{n-1}.

Note that $\mathbb{R}P^n$ has isometry group $PO(n+1)$ that admits a natural embedding $PO(n+1) \subset PU(n+1)$ as a subgroup that commutes with I. This subgroup acts linearly by $O(n+1)$ on the homogeneous coordinates $\mathbb{C}^{n+1} \setminus \{0\}$. The group $PO(n+1)$ of real isometries also fixes the quadric Q^{n-1} set-wise.

4.1. The symplectic structure on the complement of a smooth quadric

A folklore result states that the complement of a smooth quadric $Q^n \subset \mathbb{C}P^n$ is symplectomorphic to a fibre-wise convex subset of the cotangent bundle $T^*\mathbb{R}P^n$; see work by Adaloglou for a detailed proof [13, Corollary 2.4]. This claim is very closely related to the even more well-known folklore result that the affine quadric is symplectomorphic to T^*S^n; indeed, taking the double cover of $\mathbb{C}P^n$ branched along the quadric Q^{n-1} we get a projective quadric, with the complement of the branching locus equal to a subset of the double cover $T^*S^n \to T^*\mathbb{R}P^n$. Since we will need the identification to be compatible with the anti-symplectic involutions and $PO(n+1)$-actions, we reprove the existence of the symplectomorphism here.

Through each point $p \in \mathbb{R}P^n$ we can consider the element in $PU(n+1) = Isom(\mathbb{C}P^n, \omega_{FS})$ which is uniquely determined by the property that it fixes the point p, while its differential is given by multiplication by i in the tangent space $T_p\mathbb{C}P^n$. The image of $\mathbb{R}P^n$ under this linear map is thus a uniquely determined Lagrangian embedding $\mathbb{R}P^n_p \subset \mathbb{C}P^n$ of the projective plane that intersects the standard $\mathbb{R}P^n$ transversely in the point p.

The Lagrangian $\mathbb{R}P^n_0$ which passes through the origin $0 \in \mathbb{R}P^n$ in the standard affine chart is given by the imaginary projective plane $\mathbb{R}P^n_0 = i\mathbb{R}P^n$. The latter can be seen to intersect $\mathbb{R}P^n$ in the projective plane $\mathbb{R}P^{n-1}_\infty \subset \mathbb{R}P^n$ at infinity with respect to the standard affine chart. Note that

$$PO(n+1) \subset PU(n+1) = Isom(\mathbb{C}P^n, \omega_{FS})$$

acts transitively on $\mathbb{R}P^n$, as well as on the set of Lagrangian subspaces $\mathbb{R}P^n_p$ for $p \in \mathbb{R}P^n$.

Note that $\mathbb{R}P^n_0 \cap Q^{n-1}$ is equal to the Lagrangian $(n-1)$-dimensional sphere

$$\{z_1^2 + \ldots + z_n^2 = -1, \mathfrak{Re}\,\mathbf{z} = 0\} \subset \{1 + z_1^2 + \ldots + z_n^2 = 0\}$$

inside the symplectic submanifold $Q^{n-1} \subset \mathbb{C}P^n$. Furthermore, any real line through $0 \in \mathbb{R}P^n_0$ is determined by its intersection in two antipodal points on the latter sphere.

Proposition 4.1. *The open Lagrangian n-discs $B^n_p \subset \mathbb{R}P^n_p$ whose boundary is given by the $(n-1)$-dimensional sphere*

$$\partial\overline{B^n_p} = Q^{n-1} \cap \mathbb{R}P^n_p \subset \mathbb{R}P^n_p$$

are embedded and foliate $\mathbb{C}P^n \setminus Q^{n-1}$. In other words, $\mathbb{C}P^n \setminus Q^{n-1}$ is a tubular neighbourhood of $\mathbb{R}P^n$ in the form of a Lagrangian n-disc bundle on which $PO(n+1)$ acts.

Proof. The fact that the discs form an embedded tubular neighbourhood: Two discs B^n_p, B^n_q for $p \neq q$ are disjoint for the following reason. Both discs are foliated by real lines in $\mathbb{R}P^n_p$ and $\mathbb{R}P^n_q$ that pass through p and q, respectively, whose complexifications moreover have intersections with $\mathbb{R}P^n$ that again are real lines. The two real lines $\ell_p \subset \mathbb{R}P^n_p$ and $\ell_q \subset \mathbb{R}P^n_q$ passing through p and q, respectively, can be seen to have an intersection in $B^n_p \cap B^n_q$ if and only if their complexified lines coincide. Indeed, if the corresponding real lines $\mathbb{C}\ell_p \cap \mathbb{R}P^n \subset \mathbb{R}P^n$ and $\mathbb{C}\ell_q \cap \mathbb{R}P^n \subset \mathbb{R}P^n$ intersect at an imaginary point, then these lines must coincide.

Thus, if B^n_p and B^n_q intersect for $p \neq q$, we are in a situation where the complexifications of ℓ_p and ℓ_q coincide, and intersect $\mathbb{R}P^n$ in the unique real line that passes through both p and q. When this holds one can verify, e.g. by consulting the case $n = 1$, that ℓ_p and ℓ_q for two different points $q \neq p$ that live on a common line in $\mathbb{R}P^n$ intersect precisely at the two points in the generic intersection $\ell_p \cap Q^{n-1} = \ell_q \cap Q^{n-1}$. This contradiction implies that the two balls B^n_p and B^n_q must be disjoint, as sought.

The fact that the discs make up all of $\mathbb{C}P^n \setminus Q^{n-1}$: This follows from elementary topology. It suffices to note that $\mathbb{C}P^n \setminus Q^{n-1}$ is open and connected. The disc bundle constructed is also open, and its topological boundary is contained inside Q^{n-1}. □

We will now compare the symplectic structure on $\mathbb{C}P^n \setminus Q^{n-1}$ with the symplectic structure on the cotangent bundle $T^*\mathbb{R}P^n$. The Fubini–Study form can equivalently be written as

$$\omega_{FS} = \frac{i}{2}\partial\bar{\partial}\rho, \;\; \rho(\mathbf{z}) = \log\left(\frac{1 + \|\mathbf{z}\|^2}{|1 + z_1^2 + \ldots + z_n^2|}\right),$$

in the complement of Q^{n-1} in the standard affine coordinates. The function $\rho(\mathbf{z})$ extends to a well-defined potential on all of $\mathbb{C}P^n \setminus Q^{n-1}$. (It is induced by a trivialisation of the line-bundle $\mathcal{O}(2) \to \mathbb{C}P^n$ defined in the complement of Q^{n-1}.) The fact that ρ can be extended to a potential away from Q^{n-1} can also be checked explicitly, e.g. by observing that it is has the same expression in all affine charts.

Furthermore, this potential is preserved by the action of $PO(n+1)$. This can be seen by considering the homogeneous function

$$\log\left(\frac{\|\mathbf{Z}\|^2}{|Z_0^2+Z_1^2+\ldots+Z_n^2|}\right)$$

of degree zero which is invariant under $O(n+1)$ and well-defined away from $\{Z_0^2+Z_1^2+\ldots+Z_n^2=0\}$. In this manner, we get a primitive $-\frac{1}{4}d^c\rho\in\Omega^1(\mathbb{C}P^n\setminus Q^{n-1})$ of the Fubini–Study form defined on $\mathbb{C}P^n\setminus Q^{n-1}$.

Theorem 4.2. *There exists a $PO(n+1)$-equivariant embedding*

$$\Phi\colon \left(\mathbb{C}P^n\setminus Q^{n-1},-\frac{1}{4}d^c\rho\right)\hookrightarrow(T^*\mathbb{R}P^n,\theta_{\mathbb{R}P^n})$$

that preserves the primitives of the symplectic forms, which is uniquely determined by the properties:

(1) the restriction to $\mathbb{R}P^n\subset\mathbb{C}P^n$ is the canonical identification with the zero-section $0_{\mathbb{R}P^n}\subset T^*\mathbb{R}P^n$;
(2) each B_p^n is mapped into the corresponding cotangent fibre $T_p^*\mathbb{R}P^n$;
(3) the anti-symplectic involution I in the domain is identified with I_0 in the cotangent bundle, i.e. the fibre-wise scalar multiplication by -1 in the target; and
(4) when restricted to any B_p^n, the map preserves lines through the origin (as defined by the conformal structure in the domain and the linear bundle structure in the target);
(5) when restricted to any B_p^n, the map sends concentric spheres in B_p^n around the origin (defined by the Fubini-Study metric on $\mathbb{C}P^n$), to a spherical cotangent fibres of constant radius in $T_p^*\mathbb{R}P^n$ (induced by the Fubini–Study metric on $\mathbb{R}P^n$).

Proof. For the uniqueness, we note that any symplectomorphism Φ of T^*M that is the identity along the zero-section and preserves fibres must be the identity. Indeed, conjugating the symplectomorphism with a fibre-wise rescaling

$$\sigma_t^{-1}\circ\Phi\circ\sigma_t,\ \sigma_t(x)=t\cdot x,$$

yields a Hamiltonian isotopy of T^*M that fixes both the zero-section pointwise, and the Lagrangian fibres set-wise. Since such a Hamiltonian must be constant both along the zero-section as well as along the fibres, it follows that the generating Hamiltonian is globally constant. Consequently, Φ commutes with σ_t for all $t>0$. Furthermore, since $D\Phi$ must be the identity along 0_M, it follows that $\Phi=\mathrm{Id}_{T^*M}$ holds everywhere, as sought.

In order to define the $P(O+1)$-equivariant map it suffices to prescribe its restriction to a map

$$\Phi_0\colon B_0^n\hookrightarrow T_0^*\mathbb{R}P^n.$$

First note that the $O(n) \subset P(O+1)$ equivariance implies that this map is determined once we know it on the ray

$$\{y_i = 0;\ i = 2, \ldots, n\} \cap B_0^n \subset \mathfrak{Im}\mathbb{C}^n \subset \mathbb{C}^n = \mathbb{C}P^n \setminus \mathbb{C}P_\infty^{n-1}.$$

We prescribe this ray to be sent into the one-dimensional subspace consisting of this cotangent vectors that annihilate the tangent space to the embedding $\{x_1 = 0\} \subset \mathbb{R}P^{n-1} \subset \mathbb{R}P^n$ at $0 \in \mathbb{R}P^n$.

Defining the map Φ_0 with some more care along the line, we can ensure that the primitive $-\frac{1}{4}d^c\rho$ of ω_{FS} is the pull-back of $\theta_{\mathbb{R}P^n}$ under the restricted map

$$\mathbb{C}P^1 \setminus Q^0 \hookrightarrow T^*\mathbb{R}P^1 \subset (\mathbb{T}^*\mathbb{R}P^n, \theta_{\mathbb{R}P^n}),$$

where $\mathbb{C}P^1 \setminus Q^0$ is the complexified real line $\mathbb{C}\{y_i = 0,\ i = 2, \ldots, n\} \setminus Q^{n-1}$ with the quadric removed. Here the latter complexified line is endowed with the restriction of the Fubini–Study form (which again simply is the Fubini–Study form in dimension one).

In fact, since the primitive of the symplectic form is given by

$$-\frac{1}{4}d^c\rho|_{\mathbb{R}P_0^n} = \frac{1}{2}\sum_i \left(\frac{-1}{1+\|\mathbf{z}\|^2} + \frac{-1}{|1+z_1^2+\ldots+z_n^2|} \right) y_i\, dx_i = \sum_i \frac{-y_i}{|1-\|\mathbf{y}\|^4|}dx_i,$$

which when further restricted to $B_0^n = \{\|\mathbf{y}\|^2 < 1\} \subset \mathbb{R}P_0^n$ becomes

$$-\frac{1}{4}d^c\rho|_{B_0^n} = \sum_i \frac{-y_i}{1-\|\mathbf{y}\|^4}dx_i.$$

It now follows by construction that $-\frac{1}{4}d^c\rho$ is the pull-back of $\theta_{\mathbb{R}P^n}$ on all of $T\mathbb{C}P^n$ along $\mathbb{C}P^1 \setminus Q^0$ (not just on the tangent space of the latter). It thus follows that Φ is symplectic everywhere.

(1), (2), (3), (4): These points now follow immediately by construction.

(5): follows by (4) and the $PO(n+1)$-equivariance. Note that the action of $O(n) \subset PO(n+1)$ acts on B_0^n while preserving the concentric spheres around the origin. □

4.2. Definition of conical submanifolds

By using the above symplectic identification of $\mathbb{C}P^n \setminus Q^{n-1}$ with a fibre-wise convex subset of $T^*\mathbb{R}P^n$ we can make the following definition.

Definition 4.3. An immersion in $\mathbb{C}P^n$ is said to be *conical near* $\mathbb{R}P^n$ if it corresponds to an immersion in $T^*\mathbb{R}P^n$ which is conical near $0_{\mathbb{R}P^n}$ under the symplectic embedding $\mathbb{C}P^n \setminus Q^{n-1} \hookrightarrow T^*\mathbb{R}P^n$ from Theorem 4.2.

Note that immersions that are conical near $\mathbb{R}P^n$ in particular are I-invariant in some neighbourhood of $\mathbb{R}P^n$, and determined by an I-invariant transverse link in $U^*\mathbb{R}P^3$ over which the surface is a cone.

5. Proof of Theorem 1.2

We start by recalling the constructions of the standard I-invariant algebraic curves Σ_i of degrees $i = 1$ and $i = 2$. We will use parts of these for constructing our general I-invariant symplectic surfaces.

The standard $\mathbb{R}P^1 \subset \mathbb{R}P^3$ as well as the standard null-homologous unknot

$$\{x_1^2 + x_2^2 = 1, x_3 = 0\} \subset \mathbb{R}P^3$$

are both real-algebraic. The complex locus $\mathbb{C}P^1$ of $\mathbb{R}P^1$ is conical on the nose when considered inside $\mathbb{C}P^3 \setminus Q^2 \hookrightarrow T^*\mathbb{R}P^3$. Denote this surface by $\Sigma_1 = \mathbb{C}P^1$ and note that it is of degree one.

It is no longer true that the algebraic surface Σ_2 with real locus the unknot $\{x_1^2+x_2^2 = 1, x_3 = 0\}$ is conical in a neighbourhood of $\mathbb{R}P^3$; however, we can use Lemma 3.4 to Hamiltonian isotope the complex part of this algebraic knot while fixing the real locus, in order to make it conical near $\mathbb{R}P^3$. Denote the obtained conical I-invariant symplectic surface by Σ_2, which is of degree two.

Since Σ_i are conical near $\mathbb{R}P^3$, their intersections with some small tubular neighbourhood $U \subset \mathbb{C}P^3$ of $\mathbb{R}P^3 \subset \mathbb{C}P^3$ are surfaces that are conical in $T^*\mathbb{R}P^3$ under the identification in Theorem 4.2. Moreover, we can assume that the latter identification sends U to a radius-r co-disc bundle $D_r^*T^*\mathbb{R}P^3$ for some small radius $r > 0$.

In particular, $(\Sigma_i \cap U) \setminus \mathbb{R}P^3$ corresponds under Theorem 4.2 to a cone $(-\infty, a) \times T_i$ over a transverse I-invariant link $T_i \subset U^*\mathbb{R}P^3$ consisting of two components that are interchanged by I. In the case $i = 1$ each of the two components lives in the non-trivial homotopy class in $\pi_1(U^*\mathbb{R}P^3) \cong \pi_1(\mathbb{R}P^3) \cong \mathbb{Z}_2$, while in the case $i = 2$ each component lives in the trivial homotopy class.

Moreover, we will write

$$\Sigma_i^{\text{hat}} := \Sigma_i \setminus U \subset \mathbb{C}P^3$$

for the complement of the conical sub-surface $U \cap \Sigma_i \supset (\Sigma_i)_{\mathbb{R}}$. Note that Σ_i^{hat} is symplectic, embedded, I-invariant, with boundary given by the transverse link T_i.

Now consider an arbitrary smooth link $L \subset \mathbb{R}P^3$. First apply the construction in Part (2) of Example 3.2 to produce a piece of an I-invariant symplectic surface near $\mathbb{R}P^3$ that is conical near $\mathbb{R}P^3$. Denote this surface by Σ_L^0, which is conical near $\mathbb{R}P^3$ and intersects $\mathbb{R}P^3$ in the link L. Note that the boundary of Σ_L^0 can be identified with a transverse link in $U^*\mathbb{R}P^3$ that consists of an even number of components that are interchanged under I. Choosing an orientation of L trivialises this action by singling out representatives $T_L^{0,+}$ for each equivalence class, so that $\partial\Sigma_L^0 = T_L^0 = T_L^{0,+} \sqcup I(T_L^{0,+})$.

Then we concatenate (see Definition 2.8) the surface Σ_L^0 with suitable I-invariant symplectic cobordisms coming from the connected-sum construction as described in Subsection 3.2.3. We can e.g. make the choice of connecting all components of $T^{0,+}$, and then extending this connected symplectic cobordism to one which is I-invariant and consisting of two components. After suitable choices of boundary connected sums, we obtain an

I-invariant symplectic surface Σ^1_L with boundary consisting of two transverse links that are interchanged by I, has L as its fixed-point locus, and which is of genus $\operatorname{rk} H_0(L) - 1$.

The classification of I-invariant transverse knots from Proposition 2.7 implies that the boundary of Σ^1_L is transverse isotopic inside $U^*\mathbb{R}P^3$ to either T_1 or T_2 through I-invariant transverse knots, depending on its homotopy class. First adjoining the symplectic I-invariant trace cobordism produced by Lemma 2.10 and then adjoining the I-invariant "hat" produced above Σ^{hat}_i of a suitable degree $i \in \{1, 2\}$, yields a symplectic embedding of a closed connected surface of genus $\operatorname{rk} H_0(L) - 1$ that is I-invariant and intersects $\mathbb{R}P^3$ in L. This finishes the construction of a connected type-I surface of genus $\operatorname{rk} H_0(L) - 1$.

In order to construct a surface of type-I of higher genus we start by proceeding as above, constructing the I-invariant surface Σ^1_L with two transverse boundary components, and which is of genus $\operatorname{rk} H_0(L) - 1$. We then use the connected symplectic genus-k filling from Subsection 3.2.1 and double it with the involution I, after which we perform pairs of boundary connected sums as in Subsection 3.2.3 to make the resulting surface connected and invariant under I. Note that the homology class of the boundary of the surface is not altered by these operations, but that the process increased the genus of the surface Σ^1_L by $2k$. To finish, we add the I-invariant "hat" i.e. Σ^{hat}_i of a suitable degree $i \in \{1, 2\}$ in order to close up the surface.

Finally, for the construction of surfaces of type-II, the construction is similar to the case treated in the previous paragraph. In this case, however, instead of adding a pair of genus k-surfaces, we add k disjoint copies of the connected I-invariant cylinder from Subsection 3.2.2 (pushed-off to a symplectic surface), and perform k pairs of boundary connected sums with these cylinders as described in Subsection 3.2.3. Note that this process increases the genus of Σ^1_L by k. □

6. Natural questions

In future work we plan to study the following questions. First, it is very natural to consider the isotopy problem for the symplectic flexible links.

Question 6.1. Are two flexible symplectic links with the same topological invariants Hamiltonian isotopic through flexible symplectic links?

We expect that an I-equivariant version of the h-principle for high-codimension symplectic embeddings can be used to give a positive answer to the this question; see [6] for the relevant h-principle. The smooth but non-symplectic case was treated by the first author in [4].

Second, it would be useful to find a concept that is less flexible than symplectic I-invariant surfaces, but more flexible than real-algebraic curves. In the light of the constructions that were used in the present article, we make a tentative proposal of a definition.

Call an I-invariant piecewise-smooth embedded closed surface $\Sigma \subset \mathbb{C}P^3$ ***I*-invariant symplectic-conical surface with a holomorphic hat** if there exists an I-invariant neighbourhood $U \subset \mathbb{C}P^3$ of $\mathbb{R}P^3$ with contact-type boundary such that Σ is holomorphic in $\mathbb{C}P^3 \setminus U$, while it is smooth, symplectic, and conical inside U, and moreover has complex tangent planes along its intersection with $\mathbb{R}P^3$.

Question 6.2. What are the restrictions on the link type in $\mathbb{R}P^3$ obtained as the real part of an I-invariant symplectic-conical surface with a holomorphic hat, when the surface is of degree at most two?

The above question can be reformulated as a question about the existence of holomorphic hats inside $\mathbb{C}P^3 \setminus \Phi^{-1}(D^*_{<r}\mathbb{R}P^3)$, where Φ is the symplectic embedding from Theorem 4.2, where the holomorphic hat has boundary equal to a certain transverse link inside the boundary of $\Phi^{-1}(U^*_r\mathbb{R}P^3)$ of a tubular neighbourhood of $\mathbb{R}P^3$ symplectomorphic to a radius-r unit sphere bundle. Symplectic hats inside $\mathbb{C}P^2$ were studied in work by Etnyre–Golla [14]

Also note that the image of a symplectic-conical surface with a holomorphic hat under a real-algebraic knot projections to $\mathbb{C}P^2$ is an I-invariant immersions of the same type. The knot projection does not preserve the symplectic property of a submanifold in general; however, if we assume that the surface in $\mathbb{C}P^3$ is conical with complex tangents near $\mathbb{R}P^3$, its projection is an immersed symplectic-conical surface with a holomorphic hat. Indeed, the image of a (real) line under the knot projection is again a (real) line. Thus, when a surface which is conical near $\mathbb{R}P^3$ moreover has complex tangent planes along $\mathbb{R}P^3$, the cones of the aforementioned types are tangent to a complexified real lines along each of its fibres.

References

[1] J. Björklund. *Real algebraic knots of low degree.* J. Knot Theory Ramifications, **20** (2011), 1285–1309.

[2] G. Mikhalkin and S. Orevkov. *Real algebraic knots and links of small degree.* J. Knot Theory Ramifications, **25** (2016).

[3] S. Akbulut. *On algebraic structures of manifolds.* Recent advances in real algebraic geometry and quadratic forms, Contemp. Math. (Volume 155), American Mathematical Society, (1994), 11–17.

[4] J. Björklund. *Flexible isotopy classification of flexible links.* J. Knot Theory Ramifications, **25** (2016).

[5] P. Antonini, F. Cavalletti, and A. Lerario. *A proof of the Hamiltonian Thom isotopy lemma.* Note Mat., **43** (2023), 13–25.

[6] K. Cieliebak, Y. Eliashberg, and N. Mishachev. *Introduction to the h-principle.* American Mathematical Society, (2024).

[7] A. Weinstein. *Symplectic manifolds and their Lagrangian submanifolds.* Advances in Math., **6** (1971) 329–346.

[8] H. Geiges. *An introduction to contact topology*, Cambridge University Press, (2008).

[9] C. Cao, N. Gallup, K. Hayden, and J. M. Sabloff. *Topologically distinct Lagrangian and symplectic fillings.* Math. Res. Lett., **21** (2014), 85–99.

[10] V. I. Arnold. *Topological invariants of plane curves and caustics.* University Lecture Series (Volume 5), American Mathematical Society, (1994).

[11] T. Ekholm, K. Honda, and T. Kálmán. *Legendrian knots and exact Lagrangian cobordisms.* J. Eur. Math. Soc., **18** (2016), 2627–2689.

[12] G. Dimitroglou Rizell. *Legendrian ambient surgery and Legendrian contact homology.* J. Symplectic Geom., **14** (2016), 811–901.

[13] N. Adaloglou. *Uniqueness of Lagrangians in T^*RP^2*, preprint, `https://arxiv.org/abs/2201.09299`.

[14] J. B. Etnyre and M. Golla. *Symplectic hats.* Journal of Topology, **15** (2022), 2216–2269.

AVDELNINGEN FÖR ELEKTROTEKNIK, MATEMATIK OCH NATURVETENSKAP, HÖGSKOLAN I GÄVLE, SE-801 76 GÄVLE, SWEDEN
Email address: johan.bjorklund@hig.se

DEPARTMENT OF MATHEMATICS, UPPSALA UNIVERSITY, BOX 480, SE-751 06 UPPSALA, SWEDEN
Email address: georgios.dimitroglou@math.uu.se

Proceedings of 29th Gökova
Geometry-Topology Conference
pp. 57 – 58

Corks

Selman Akbulut

ABSTRACT. Remarks relating the various notions of corks.

I recall, my advisor Kirby kindly named the object I had constructed in [A2] "Akbulut cork". Since then, this concept went through some evolutions, resulting some confusions between its various different definitions. To clarify them, I decided to collect here mutually consistent definitions.

Definition: A **loose cork** of a smooth manifold M is a pair (W, f), where W is a smooth contractible submanifold of M with involution $f : \partial W \to \partial W$, such that cutting W out of M then regluing with f changes the smooth structure of M. A **cork** (W, f) is a loose cork, which is a Stein manifold.

We call the operation of cutting W from M and regluing by the map f *a loose cork-twisting operation on* M, or simply cork-twisting operation on M. At first glance the "Stein" condition might look artificial, but it really is not, because by [AM2] every loose cork (W, f) contains a cork, hence cork is a smaller fundamental object. Furthermore, given a contractible manifold with involution on its boundary (W, f), it is a difficult to decide whether W is a loose cork (is there a corresponding M?), or does f extend to a self diffeomorphism of W?. To answer the second question in the negative, it suffices to find a pair of knots $\gamma, \gamma' \subset \partial W$ such that γ is not slice but γ' is slice in W, with $f(\gamma) = \gamma'$. As in the example of Figure 9.1 of [A1] (if γ was slice you would get a violation to the "Adjunction Inequality for Stein manifolds", which had been proven in [AM1]).

To find such examples (W, γ, γ'), symplectic topology can be useful. For example, when W is a Stein manifold, attaching W a 2-handle along a knot $\gamma \subset W$ with $TB(\gamma) - 1$ framing extends the Stein structure from W to $W \smile h^2_\gamma$ (In fact, by [LM] and [AO], we can even compactify any Stein manifold into a closed symplectic manifold which could play the role of M in the above definition of cork). Then by the Adjunction inequality proved for the Stein manifolds in [AM1], applied to $W \smile h^2_\gamma$, shows that γ can not be slice.

Sometimes after I wrote [A2], I realized that, even though the cork, I constructed there, looks like the manifold which Mazur had defined in his Annals paper, it is not. Mazur's manifold is not even a "loose cork"!

Matveyev [M] generalized [A2] by proving any exotic copy M' of a simply connected closed smooth 4-manifold M can be obtained by performing a loose cork-twisting operation on M. Hence by [AM2] it can be obtained by a cork-twisting operation on M. Curtis-Freedman-Hsiang-Stong [CFHS] proved a weaker version of [M] without mentioning a cork, namely "M' can be obtained by cutting out a smooth compact contractible submanifold of M, then gluing another smooth compact contractible manifold with the same boundary".

By weakening the involution condition f on a loose cork (W, f) one can construct an infinite order loose cork; [A3] and [G] are such examples. Notice that in minute 1:14:20 of my talk in [A4] I am trying to construct an infinite order cork (with "Stein" condition), which we still don't know if exists.

References

[A1] S. Akbulut, 4-*Manifolds*. Oxford University Press, (2016).

[A2] S. Akbulut, *A Fake compact contractible 4-manifold*. Journ. of Diff. Geom., **33** (1991), 335-356.

[A3] S. Akbulut, *On infinite order corks* PGGT (2017). IP Paperback, ISBN 9781571463401 151-157.

[A4] S. Akbulut, *Handlebodies of exotic 4-manifolds*. (A talk on infinite order corks), FRG conference, SCGP-2014 Part 3/3: `https://www.youtube.com/watch?v=J0EaxAyODTo&list=PLZQgOF7a2hhVpsnYRyMVZXALiVUpB1Dk6&index=3&t=4s`

[AM1] S. Akbulut and R.Matveyev *Exotic structures and adjunction inequality*. Turkish Journal of Math **21** (1997) 47-53

[AM2] S. Akbulut and R. Matveyev, *A convex decomposition theorem for 4-manifolds*. IMRN International Math Research Notices, **(7)** (1998), 371-381

[AO] S. Akbulut and B. Ozbağcı, *On the Topology of Compact Stein Surfaces*. IMRN, **15** (2002), 769-782

[AY] S. Akbulut and K. Yasui, *Corks, plugs and exotic structures*. J. Gokova Geom. Topol., **2** (2008), 40–82.

[G] R. E. Gompf *Infinite order corks*. (2016) arXiv:1603.05090.

[LM] P. Lisca and G. Matic *Tight contact structures and Seiberg-Witten invariants*. Invent. Math., **129(3)** (1997), 509–525.

[M] R. Matveyev, *A decomposition of smooth simply-connected h-cobordant 4-manifolds*. J. Differential Geom., **44(3)** (1996), 571–582.

[CFHS] C.L. Curtis, M.H. Freedman, W.C. Hsiang, R. Stong, *A decomposition theorem for h-cobordant smooth simply-connected compact 4-manifolds*. Invent. Math., **123(2)** 123 (1996), 343–348.

Gökova Geometry Topology Institute, Muğla, Turkiye

Email address: `akbulut.selman@gmail.com`

Scenes from Gökova/2024

Mud makes health.

This year's hero is explaining some Physics.

Making friends with the Cleopatra Beach roosters.

29th GÖKOVA GEOMETRY-TOPOLOGY CONFERENCE
May 27-May 31, 2024

PARTICIPANTS

Akbulut, Selman	GGTI, Türkiye
Akyar, Turgay	Middle East Technical University, Türkiye
Akyol, Özgür Deniz	Middle East Technical University, Türkiye
Alabay, Hilal	Bursa Technical University, Türkiye
Arıkan, Fırat	Middle East Technical University, Türkiye
Atalan, Ferihe	Atılım University, Türkiye
Bhupal, Mohan	Middle East Technical University, Türkiye
Ceylan, Berk	Türkiye
Degtyarev, Alexander	Bilkent University, Türkiye
Demory, Aloïs Pierre Marc	Sorbonne University, France
Dimitroglou Rizell, Georgios	Uppsala University, Sweden
Duman, Okan	Yıldız Technical University, Türkiye
Erdal, Mehmet Akif	Yeditepe University, Türkiye
Erdal, Esma Dirican	FMV Işık University, Türkiye
Finashin, Sergey	Middle East Technical University, Türkiye
Geven, Berat	Middle East Technical University, Türkiye
Golovko, Roman	Charles University, Czech Republic
Hind, Richard	University of Notre Dame, USA
Itenberg, Ilia	Sorbonne University, France
Karaca, Emel	Ankara Hacı Bayram Veli University, Türkiye
Kayakökü, Mehmet Mustafa	Middle East Technical University, Türkiye
Kişisel, Ali Ulaş Özgür	Middle East Technical University, Türkiye
Kule, Memet	Kilis 7 Aralık University, Türkiye
Mikhalkin, Grigory	Université de Genève, Switzerland
Mursalzade, Aysel	Middle East Technical University, Türkiye
Önder, Turgut	Middle East Technical University, Türkiye
Orcan, Yeda Nazlı	İstanbul Bilgi University, Türkiye
Özkan, Ayşenur	Kırıkkale Üniversitesi, Türkiye
Petrunin, Anton	Pennsylvania State University, USA
Rozhkovskaya, Natasha	Kansas State University, USA
Shinder, Evgeny	University of Sheffield, England
Shkolnikov, Mikhail	Bulgarian Academy of Science, Bulgaria
Souri, Marzieh	IPM, Iran
Suluyer, Hasan	Middle East Technical University, Türkiye
Şen, Muhammed	Middle East Technical University, Türkiye

Taşpınar, İrem Özge	Middle East Technical University, Türkiye
Toussaint, Antoine	Université de Genève, Switzerland
Ulusay, Selin	Bilkent University, Türkiye
Yıldız, Eylem Zeliha	Duke University, USA
Yıldırım, Yasemin	Middle East Technical University, Türkiye
Yokuş, Fatih	Middle East Technical University, Türkiye
Zharkov, Ilia	Kansas State University, USA
Zvonkine, Dimitri	Université De Versailles Saint-Quentin-En-Yvelines, France